T0234310

Springer Textbooks in Earth Sciences, Geography and Environment

The Springer Textbooks series publishes a broad portfolio of textbooks on Earth Sciences, Geography and Environmental Science. Springer textbooks provide comprehensive introductions as well as in-depth knowledge for advanced studies. A clear, reader-friendly layout and features such as end-of-chapter summaries, work examples, exercises, and glossaries help the reader to access the subject. Springer textbooks are essential for students, researchers and applied scientists.

Christopher Gomez

Point Cloud Technologies for Geomorphologists

From Data Acquisition to Processing

Springer

Christopher Gomez
Kobe University
Kobe, Hyogo, Japan

ISSN 2510-1307   ISSN 2510-1315   (electronic)
Springer Textbooks in Earth Sciences, Geography and Environment
ISBN 978-3-031-10977-5   ISBN 978-3-031-10975-1   (eBook)
https://doi.org/10.1007/978-3-031-10975-1

To my wife, Junko, for her continuous support,

and to my parents and brothers who have been the silent victims of my scientific tribulations that seldom brings me back to the country road.

Preface

Geomorphology is the discipline that investigates the surface of the Earth, by examining both the landforms (literally the land–forms) and the processes that shape those landforms. The landforms are associated with processes that can be endogenous (such as earthquakes and volcanic activity) and exogenous (such as rainfall erosion and wind deflation), as well as any combination of the two (e.g. landslides, debris flows and other mass movements that are often at the crossroad between slopes generated by tectonic uplift for instance and the role of water mixed with sediments under the action of gravity, etc.). Other processes notably include human influences and zoogeomorphology, for instance, etc.

To measure these processes and landforms, geomorphologists use a wide variety of tools, among which we can now count pointcloud technologies, which have become a household name in most laboratories. In this book, which is also designed as a manual, I have tried to give a presentation of the different methods to collect field data and then process them for geomorphological purposes. The first group of chapters (more technical) is then completed by a set of chapters divided between geomorphological environments and processes. The difficulty there, was to give an introduction to some concepts that were of importance, but without it becoming yet another introductory geomorphology book. Choices were thus made and for instance, in the chapter on landslide, I give some elements of slope stability and soil mechanics, because in combination with point-cloud technologies, they can—I believe—bring to the discussion table the importance of some of the topographic details that were estimated or just overlooked.

As I started my geomorphologist career at Paris 1 Sorbonne University in France, before moving to Paris 7 University (now Paris University), digital elevation models and gridded data were just emerging in the field and the geospatial education in GIS and other applied techniques was still in its infancy, and it certainly took some extra work to get up to speed and remaining relevant through my career, which is a bit more than a decade after my Ph.D. when I am writing those lines. And I have certainly felt the thrill of these new techniques and seeing growing in the field of geomorphology all the new options that were offered to us, but at the same time, I did not take much time to reflect on the real nature of the changes, metamorphosis and modifications that new technologies were operating on the scientific field. Of course, I would rather like to see the tools serving the idea, but I am forced to admit that in the present case the tool has provided deep changes to the very approach and the vision that we have of geomorphology. This is why I am keeping Chap. 7 to discuss some of those points, although I already provide some quick hints in the introduction.

When I wrote this book, I had the students of geomorphology in mind, and although I know that they can find several excellent monographs on landforms and the Earth surface processes, I have failed to identify a complete manuscript on pointcloud technologies that have recently geomorphology. Of course, there are several books in remote sensing, in civil and other engineering fields, but I did not find anything that was easily accessible to all of their physics and mathematics background. I have therefore Furthermore, I certainly had my own students in mind, thinking that it would be nice to have a short-handbook that they could go to when they need to either collect or process pointclouds in a given geomorphological environment and at a given scale.

As any monograph, it is a work in progress and even after publication, we will all find different ways to improve the manuscript further, and in advance, I thank the readers for any constructive comments they will provide me with.

Kobe, Japan   Christopher Gomez
2021

Contents

Abbreviations

AGAST	Adaptive and generic accelerated segment test
ALS	Aerial laser scanner
ASIFT	Affine scale-invariant feature transform
AST	Accelerated segment test
DEM	Digital elevation model
DSM	Digital surface model
FAST	Features from accelerated segment test
GCP	Ground control points
GNSS	Global Navigation Satellite System
HRT	High-resolution topography
LiDAR	Light detection and ranging
M3C2	Multiscale model-to-model cloud comparison
MAE	Mean absolute error
MBE	Mean bias error
MOPS	Multi-scale oriented patches
PPK	Post-processing kinematic
RMSE	Root mean square error
RTK	Real-time kinematic
SDE	Standard deviation of error
SE	Standard error
SfM-MVS	Structure from motion–multiple-view stereophotogrammetry
SIFT	Scale-invariant feature transform
TIN	Triangulated irregular network
TLS	Terrestrial laser scanner
TOF	Time of flight
UAV	Unmanned autonomous vehicle

Pointcloud and Geomorphology—Introduction

1

Abstract

The first chapter is an introduction to the book. It explains what is meant by pointcloud through the manuscript as well as how the technology has emerged through the field of geomorphology. Within this framework, the laser and photogrammetric technologies (airborne laser scanner, terrestrial laser scanner and structure from motion multiple view stereophototogrammetry) used through the book are briefly introduced, as well as how these methods have come to combine with and complete traditional geodesy. The first chapter will also be the stepping stone to the discussion, which is further detailed in Chap. 7. It explains how these new technologies need to be accompanied with paradigm modifications and how the single-point measurement to the point-cloud has generated a shift in how geomorphologists work in the field and in the virtual space. From this introduction–that includes both the technological aspects and the conceptual aspects further developed in the book–the chapter ends with the structure of the other chapters and how they logically articulate with one another.

Learning Outcomes: Data Acquisition, Processing and the Logic of this Book

- After reading this chapter, you should have an understanding of pointcloud acquisition and processing concepts.
- You will know the differences between existing structured topographic data such as DEMs and DSMs and the unstructured point-cloud.
- You will know how pointclouds have transformed measurement approaches in geomorphology from a technical perspective (conceptual frameworks are discussed in the last chapter)

Finally, you should have a clear view of the overall logic and content of the book.

Chapter Content and Objectives

This chapter presents the topic the book is dealing with, it provides some overviews to the linkages outside geomorphology as well, and it also presents some questions that are pertinent to the importance of pointcloud technology in the evolution of geomorphological sciences. The question is extended by an example in soil mechanics, and once the reader will have gone through the more technical Chaps. 2–6 , it is in Chap. 7 that I will invite you to revisit this example, but this first chapter plants a few seeds, which will help through the development and the discussion.

- Express what pointclouds in geomorphology are and how they differ from other techniques.
- Present the structure of this book.

A Chapter for Geomorphologists Needing an Introduction to Point-Cloud Technologies

This chapter is for geomorphologists or those who want to become geomorphologists and are interested in learning the basics of pointcloud technology. The first chapter is setting the scene and then is shaped as a guide to help readers find the information they are interested in, in the most effective way. It is then a short guide for the reader to the "technical part" (Chaps. 2 and 3), the application chapters (Chaps. 4–6) and the discussion (Chap. 7). In other words, if the reader is looking for an explanation on how to acquire pointcloud (Chap. 2), or how to process it (Chap. 3), or examples in the field of geomorphology, then the reader may not need to spend reading through the all book. A good reader is an effective one.

Introduction

The present book deals with pointclouds (or point-clouds), which can be defined as points that are numerous enough so that they cannot be realistically counted, and we refer to them as pointclouds. In Japan for instance, the term of "point swarm" or "point flock" ("tengun") is used and reflects well this concept of a whole including too many parts, so that they cannot be counted. Pointclouds can describe numerous types of objects (including statistical series, it does not have to be a physical objects), but in the present book, we concentrate on land surface, the "main-beef" of geomorphologists, but we will allow ourselves to interact a few times with the vegetation as well, as we sometimes need to eliminate it to "see through" and extract the topography; in other cases, the vegetation will be extracted and used for it reveals details about the geomorphology as well.

As for geomorphologists, or users of geomorphology (such as geologists and civil engineers or archaeologists sometimes), the measure and reconstruction of Earth and other planets' landforms is nowadays a routine scientific operation. If this operation isn't new, methods have been progressively changing. Especially, the most recent progresses in both "field-acquisition" tools and computation have brought new sets of techniques and methods to produce and process pointclouds.

Pointcloud in geomorphology often results in large sets of points which can be spatially very dense, where each point is difficult to identify from the other, unless a zoom is performed. Consequently, pointclouds are often associated with high-resolution topography, which is one of the areas of research that arisen in recent decades and spread to most Earth science sub-disciplines. Pointcloud HRT is arguably the product of the progresses made in geodesy, which allows the measurement of the Earth from the planetary scale to sub-millimetre scales, within a few millimetre precision in z, with the multiplication of measures down to a few millimetre steps in x and y. Among the variety of techniques used to measure the surface of the Earth to create HRT, pointclouds occupy a specific niche as they provide a cloud of points with the positions in 2D or 3D, in combination with parameters attached to each point, such as the altitude, the *RGB*. values, the return intensity for lasers, the number of returns for lasers, so that each point has a format that can be expressed as $X[x, y, z, R, G, B, N, I, …]$.

This contrasts with the traditional data source the geomorphologists have used: the digital elevation model (DEM) and the digital surface model (DSM)—please note that there are other spatial products (topographic maps, etc.), because the two are respectively mathematical representation of the bare Earth and the Earth plus any elements on the surface. These are the most popular forms of mathematical representations of elevation data, and they are usually recorded in a grid format so that the altitude of each cell Z can be formulated as Eq. 1.1:

$$\sum_{i=1}^{m}\sum_{j=1}^{n} Z_{i,j} \qquad (1.1)$$

where the grid on a Cartesian coordinate is perpendicular on m, n. The two types of datasets (unstructured pointclouds and gridded DEM/DSMs) are not incompatible; however, very often the DEMs and DSMs are derived from pointclouds.

Such a grid can emerge from an original point-cloud P that is a set (Eq. 1.2):

$$\sum_{i=1}^{n} P_i\{x, y, z, i\}, \qquad (1.2)$$

from which x, y, z are the attributes of location in a 3D Cartesian space, and i represents the attribute attached to the points. For instance, if the i value records the return number for an ALS dataset, the first return could be on vegetation for instance, and the last return could be considered as being on the ground. It can be a returned intensity that corresponds to the ground for instance, which then can be simply filtered by defining a filtered point-cloud, so that P^f is defined as all the points where the characteristics $\{i\}$ = "last return" (Eq. 1.3):

$$\sum_{j=1}^{\sum_{i=1}^{n} P_{i\{i\}}=\text{last}} P_j^f\{x, y, z\} = \sum_{i=1}^{n} P_i\{x, y, z, i = \text{last}\} \qquad (1.3)$$

The last return point-cloud can then be informed to be the ground through a basic filtering process, simply based on individual characteristics of points. Such filtering can also be accomplished from the morphology of the point-cloud and from spatial statistics, which helps filter the ground out of vegetation and any other reflector (for instance, it is possible to filter out points that are less than a given spatial density).

In other words, pointclouds are a spatial entity that allows the practitioner, the researcher, etc., to extract the characteristics and morphology of a spatial object. This occurs in a two steps fashion: (1) acquisition of the pointcloud and (2) processing of the pointcloud.

The tools generating such pointclouds for topographic purposes are the fixed terrestrial laser scanner (TLS) and mobile TLS as well as the airborne laser scanner (ALS)—the methods of acquisition are developed in Chap. 2. Although those technologies are rather "new", they have been rapidly integrated by geoscientists. In 1999, Ackermann was beginning one of his articles in the journal Photogrammetry and Remote Sensing by "Airborne laser scanning represents a new and independent technology for the highly automated generation of digital terrain models (DTM) and surface models. Its development goes back to the 1970s and 1980s, with an early NASA system and other attempts in USA and Canada" [1].

The research article by Ackermann in 1999 was one of the first ones on LiDAR for geomorphology and geodesy and about 20 years later the technology that was described as "new" then is now part of most curriculums of physical geography and geomorphology. ALS has become a common place in the toolbox for research in tectonic geomorphology [13]. Fault lines movements that were not visible from Shuttle Radar Topography Mission can now appear clearly to the naked eyes thanks to 1-m resolution LiDAR data (Fig. 1 in [13]). And today, the majority of HRT is created from LiDAR data [5]. As it is now possible to capture full surfaces instead of a few selected measurements, scientists have also started to work in 4D (3D over time) to monitor landslides for instance [8] and to try understanding how surfaces evolve instead of having one measurement that characterizes a surface. It is therefore a paradigm change, as the thought process has changed thanks to the tools and methods evolution.

Furthermore, pointclouds have critically expanded the traditional vertical maps to the possibility to measure out-crops and other irregularities that could not be accounted for before. For instance, TLS are oblique views on the land-scape, so that they can also capture outcrops and all sub-vertical surfaces that are not well sampled by ALS [16]. In their work, Young et al. [16] compared the use of ALS against TLS to image seacliff erosion in California. At Solana Beach, the cliff displays a cave of more than 8 m depth and less than 4 m height, so that such shape can only be characterized using a TLS, and ALS would miss the details of such a feature. Comparing further the ALS and TLS datasets, the author has shown that erosion of the cliffs could go totally undetected using ALS. Overall, ALS tended to only detect large areas of erosion (this is to be explained to the incident angle of the laser to the sub-vertical cliff, which needs sufficient surface change to be "seen" from ALS), missed smaller ones, and ALS tended to also under-estimate the size of the eroded zones it detected. Conse-quently, TLS is now omnipresent in most of the field-based sub-divisions of geomorphology. It can be used as a one off characterization of an object or a surface, or it can be used as a temporal series.

In 2020, when I am writing the manuscript, laser tech-nologies remain relatively expensive, and this has con-tributed to the success of another technology that generates point-clouds: Structure from motion and multiple view stereophotogrammetry (SfM-MVS), especially because it can be collected from low-cost platforms such as unmanned autonomous vehicle (UAV) (e.g. Westoby et al. 15). In turn, low-cost UAVs, spearheaded and mostly represented by the Chinese maker DJi® have further helped this development process.

The apparent technological change and progress is actu-ally hiding a deeper change in the field of geomorphology itself and now the geomorphologists engage with their field data (we will discuss this element in the last chapter).

1.1 Adjustments Needed in the Working Paradigms of Geomorphologists

1.1.1 From Single Measurement to Pointcloud

A pointcloud is an ensemble of points that are far too numerous to be counted or appraised as a single unit, and the unit we can visually recognize and comprehend is the cloud, or the group of points. First, it is important that a pointcloud does not really exist; it is just based on a limitation of the human brain and computing abilities. Most humans can instantaneously remember and count up to 6 elements (just try and throw in front of you any set of 6 objects on a table, you will see that you should be able to count those by just looking at them, and so in an instant). If you are talented, you may be able to count 8 objects in a blink of an eye. But what happens when you are looking at 1000 points, or 1,367,126? Unless you are some sort of Hollywood movie genius, it may take you some time to count those, and even further if you have to deal with individual characteristics of each object. This is when our brain stops looking at each individual but rather try to find patterns and group the ele-ments following those patterns. For example, if you are looking at a pointcloud of a geomorphological object, what do you see? Do you see the points or is your brain already forming an image of the object?

This somewhat benign difference between a point and a pointcloud is actually the source of a significant shift in the mental approach to the geomorphological object. Indeed, a point is one chosen measure of a feature, such as its top, or a wall, the position and altitude of a rock, and if you are a bit more patient, a transect in a valley made of a series of points. However, when the geomorphologist is now using a laser scanner or a photogrammetric method (we will discuss those in the next chapters), we are not making a conscious choice of choosing a point. For instance, the terrestrial laser scanner has a mechanical head that rotates horizontally and vertically and records the laser return from the measured object. As a geomorphologist, you are not choosing where the point of measure will be, you are leaving this part to the machine, and you will attempt to collect the maximum number of measurements from which, during processing, you may then decide to extract one or a set of parameters and measures from the point-cloud. You are not making a choice of a measurement in the field based on a set of assumptions anymore, you are bringing back to a virtual space a repre-sentation of the geomorphology, and you perform mea-surements, not in the field anymore but in a virtual environment. Pointcloud technologies have pushed the geomorphologists from being field scientists to virtual environment scientists. You may then answer me that we are still going to the field with laser scanners or UAVs from which we take photographs, but the scientific process and the geomorphological reflection do not have to occur and do not occur in the field anymore.

1.1.2 From Error Measurement to Error Estimation

The transformation from one or a manageable number of measurements at the human-scale to a pointcloud where each individual measurement can only be handled by a machine also has implications on how scientists and engineers have to deal with the error. The error at the point level escapes its point but becomes part of a group, part of the cloud, from which indicators such as the root mean square error (RMSE) are extracted. Of course, it is not impossible to have a point-level estimation of error, but it is beyond the time-frame and eventually the ability of today's tools. For instance, let's imagine that you are calculating the error of a given measurement based on another set of measurements. For a single point, you can compare a GNSS geodetical point compared to a post-processed PPK-GNSS point, with let us say several months of recording. This way you have a base point, that is maybe precise to the millimetre in X, Y and Z. In this case, you have a single point and you place your apparatus to collect a second point at the exact same position, and then you can get a value of your error. Now, if you are creating a point-cloud and comparing two point-clouds, and let's imagine that your pointclouds have a high precision and accuracy to a centimetre or to a millimetre. How can you compare the error between the two sets? First, you need to consider the fact that regardless of how dense are your measured pointclouds, you will certainly not have all the points overlapping. Because you cannot control where the laser hits the target and because you cannot control where the points from the photogrammetric method (let us say for instance the popular structure-from-motion and multiple-view-photogrammetry), how can you measure the error between the two pointclouds? One of the methods is to create an interpolated surface on which your second point-cloud is projected, but you are not comparing two measurements anymore, you need to interpolate the data. Whether your dataset is at a metre scale, or a millimetre scale, in both cases there is a new factor of uncertainty, and you are not in a position to purely measure error, it needs to be estimated from various statistical and geometric transformations. Now, imagine you invested in a very expensive laser scanner and you are looking at a sandy granular surface, and although looking at the imbrications of the grains on the surface might be your objective, how true is your dataset? If you are interpolating this highly rough dataset to measure the error, how true and representative the error is and the measurement itself is really difficult to say. Of course, you can compare different statistical indicators and surface indicators, but they fall into the same "object scale" or statistical level, and at the point-level, as you don't know what you are measuring anymore, you also don't know what error you are measuring either. Of course, conventions and indicators generated and proposed in this book are supposed to be representative enough, but if you scratch the concepts and ideas, there are more questions than answers that may be generated by this approach in geomorphology.

1.1.3 The Need for a Paradigm Shift: Shaping the Question

On top of this inherent change to the geomorphologists' datasets and how they interact with it, there remains a paradigm shift that needs to be accomplished. This paradigm shift lies in the field of process geomorphology and the measure of processes from deposits.

Let's go back in time a little bit, when I was still a student and we were building topographic profiles from total stations and other topographic tools. Let's say we had a landslide of a couple of tens metres wide and a couple of hundred metres long, and we measured through some different interpolation techniques the volume of the deposit and the volume of the landslide scar as well as the "hole" left on the mountain slope. Because of the types of measurements and the need to interpolate the measurements with a lot of uncertainty, we knew that we would measure erosion and transported material with an error in the range of several cubic to tens of cubic metres if not more. And "this was ok".

Today, we have access to pointcloud technology and we can scan the all landslide surface as well as the whole deposit, and if we had a pre-landslide scan, you can even measure the volume of sediments transported from the landslide from its deposit, or even be tempted to compare the size of the hole from the landslide and the deposit. If the deposit is larger than the hole itself, you may then define how much erosion and material incorporation occurred during the landslide movement until it stopped.

As a geomorphologist, this is the kind of measurement you could make to understand the processes that may have taken place in a valley for instance. It is certainly a perfectly valid method and train of thoughts, and I have done similar calculations for volcanic geomorphology for instance, but I would also argue that if you use HRT pointclouds, then this thought process is not valid anymore.

Why is that so?

Imagine that you have a rock slope, very compact. Then occurs a large mass movement that partly grinds the material. For the sake of the argument, let's agree that in this thought experiment, there is the absence of any erosion or deposition during the mass movement displacement. Now, if you compare a pre-event dataset and a post-event dataset (using the common difference of DEM method or DoD), it is most likely that the cavity left on the wall where the mass movement started will be much smaller than the deposit itself. This is due to the fact that the mass movement is now a mixture of gas (air) and angular particles of different shapes that will appear to be a much bigger bulk volume than the original space in the wall. Once again, let's imagine that the grinding of the material is proportional to the travelled distance. The same rock, decompressed and broken into large angular rocks may then be of a higher bulk volume than the further grinded and rounded blocks of a deposit that could have ran further downstream. And, now if you imagine that your mass movement starts from a very elevated surface and that the material, while flowing for kilometres, is becoming smaller and smaller in size, then the deposit volume will appear slightly smaller than the very angular rocks it was made of, and if the deposit surface roughness is different, if the deposit is more elongated and less thick, then from a technical perspective as well, your deposit will display another apparent bulk volume. A geomorphological method that was acceptable with traditional data becomes less acceptable when using HRT data.

If we go back to my time-travelling machine, when those processes were measured from a few sparse measurements combined with some sketch and equations to characterize the geometry of the deposit, then it is most likely that the decompression-induced change or fracturation-induced bulk volume change would disappear in the error margin. But, if you use HRT pointclouds, you can make measurements to such a level of precision and accuracy of the surfaces that calculating volumes and volume of erosion, deposition, transport…etc. is not significant anymore. Making a perfect millimetre precision and accuracy dream measurement may actually bring nothing to the quality of your erosion measurement and calculation, because what you are measuring is not what you think you are measuring. You are measuring erosion, deposition and a wide range of other associated phenomena, including decompression, fracturation, distance-related grinding, etc.

If this needs to be integrated in geomorphology, there is, however, no necessity to reinvent the wheel as it is a problem that civil engineers have been dealing with, when working with building and infrastructures notably. It also means that if you want an accurate measurement of erosion or deposition, the otherwise easy pointcloud acquisition techniques may need to be completed with laboratory experiments or field measurements of the size of the gaps in a deposit, or the blocks, etc. And there is certainly a field of investigation in geomorphology that needs to be opened, in

order to be able to have methods to evaluate the change that relates volume to bulk mass.

1.2 Logic and Content of this Book

I created this book, like most books that have been written I believe, not as an exhaustive compendium on the topic (I don't think that this is humanly possible anymore looking at the breadth of material published in science in general), but as a manual for undergraduate and graduate students of geomorphology who would like to start using pointcloud technologies, as well as an introduction manual for the scientist who would like to develop his skillset in the field of pointcloud technologies. The book therefore starts with a presentation of some of the main tools and methods using laser technologies and structure from motion and multiple view stereophotogrammetry (SfM-MVS) to acquire the dataset in Chap. 2. Once we will have an understanding of how to collect and create data, the next chapter (Chap. 3) will introduce you with different methods and tools to process data and extract different derivatives. In this section, we will also use the LidR package in R to explain how to extract features and how to filter your dataset for vegetation and other elements that are located on the surface.

These two first chapters are very general and are an introduction to the use of pointcloud technologies, but they are not manual of photogrammetry or laser technologies, they take the approach for a reader who is a geomorphologist, a physical geographer, a geologist, who may not want to become a technical specialist, but needs the tools and methods to answer questions that are pertinent to geosciences. And at present, there are still very little resources for us geoscientists and geomorphologists as most of the books and manuals are mostly developed for engineers and students who want to become technicians of these methods.

Logically, from these two first chapters, then stems a set of chapters on different geomorphological features and groups: floodplain and coastal geomorphology (Chap. 4), mountain and slope geomorphology (Chap. 5), volcanic geomorphology (Chap. 6), before ending with a discussion chapter (Chap. 7), which echoes and build on some of the concerns developed in this introductory chapter.

When this book was generated, it was with the idea that it may not be necessary to read it in a linear fashion and that the reader may want to go to some specific information. On top of a content by chapter at the start of the book, there is also a table examples and tools. You can thus browse through the table of examples (for which there are a few lines of explanations each time) and then just go to the example that interests you. Most of the worked examples come with the corresponding lines of code in R or the "how-to" of the software that has been used.

Conclusion—What Is Ahead for Geomorphologists

Arguably, the development of laser and photogrammetric technologies is not the end of the road for geomorphologists. There are numerous areas where progress needs making. First, the capture technology has made tremendous strides, but not the processing of the data nor the way we think about surfaces. During the emerging phase of pointcloud technologies, Ackermann [1] had stated that airborne laser scanning was a new technology for the generation of DSMs. This vision, although true and essential, discards the real 3D nature of pointclouds to reduce it back to a 2.5D dataset. At the time, it was seen as an improvement of existing methods, but it has now appeared that it holds other opportunities to apprehend surfaces. For instance, this issue has started to be tackled by Brodu and Lague [2]. But algorithms could notably be improved by stopping to look at the point-clouds as "points", but for what they are, the extremity of a vector between a laser or a camera pin-hole and a target, meaning that the laser travels through a space that can be defined and eventually defined as empty or semi-empty. The use of multi-wavelength lasers can also identify object types and rock compositions, and although still in their infancy, the potential for data construction and information extraction is still largely untapped.

The other pendant of the data format transformation and this movement from 2D to 2.5D then 3D and eventually change over time creating 4D dataset are the necessity to develop adapted data representation and software to build and explain the measures that are made. There is also a need to develop algorithms that take advantage of these data developments. Recently, with the new version of QGIS for instance, there is a full integration of 3D *.las dataset into the GIS environment, but compared to other 2D and 2.5D datasets, there is still a lack of algorithms to process the data into meaningful information.

References and Suggested Readings

1. Ackermann F (1999) Airborne laser scanning-present status and future expectations. ISPRS J Photogramm Remote Sens 54:64–67
2. Brodu N, Lague D (2012) 3D terrestrial LiDAR data classification of complex natural scenes using a multi-scale dimensionality criterion: applications in geomorphology. ISPRS J Photogramm Remote Sens 68. https://doi.org/10.1016/j.isprsjprs.2012.01.006
3. Cwiakala P, Gruszczynski W, Stoch T, Puniach E, Mrocheri D, Matwij W, Matwij K, Nedzka M, Sopata P, Wojcik A (2020) UAV applications for determination of land deformations caused by underground mining. Remote Sens 12:1733. https://doi.org/10.3390/rs12111733
4. Delong SB, Youberg AM, DeLong WM, Murphy BP (2018) Post-wildfire landscape change and erosional processes from repeat terrestrial lidar in a steep headwater catchment, Chiricahua Mountains, Arizona, USA. Geomorphology 300:13–30
5. Glennie CL, Carter WE, Shrestha RL, Dietrich WE (2013) Geodetic imaging with airborne LiDAR: the Earth's surface revealed. Rep Prog Phys 76(086801):1–24
6. Gomez C, Shinohara Y, Tsunetaka H, Hotta N, Bradak B, Sakai Y (2021) Twenty-five years of geomorphological evolution in the Gokurakudani Gully (Unzen Volcano): topography, subsurface geophysics and sediment analysis. Geosciences 11:457. https://doi.org/10.3390/geosciences11110457
7. Goodwin NR, Armston JD, Muir J, Stiller I (2017) Monitoring gully change: a comparison of airborne and terrestrial laser scanning using a case study from Aratula, Queensland. Geomorphology 282:195–208
8. Jaboyedoff M, Oppikofer T, Abellan A, Derron M-H, Loye A, Metzger R, Pedrazzini A (2012) Use of LiDAR in landslide investigations: a review. Nat Hazards 61:5–28
9. Jones LK, Kyle PR, Oppenheimer C, Frechette JD, Okal MH (2015) Terrestrial laser scanning observations of geomorphic changes and varying lava lake levels at Erebus volcano, Antarctica. J Volcanol Geoth Res 295:43–54
10. Kociuba W, Kubisz W, Zagorski P (2014) Use of terrestrial laser scanning (TLS) for monitoring and modelling of geomorphic processes and phenomena at a small and medium spatial scale in Polar environment (Scott River—Spitsbergen). Geomorphology 212:84–96
11. Leroueil S, Kabbaj M, Tavenas F, Bouchard R (1985) Stress-strain-strain rate relation for the compressibility of sensitive natural clays. Geotechnique 35:159–180
12. Lin Z, Kaneda H, Mukoyama S, Asada N, Chiba T (2013) Detection of subtle tectonic-geomorphic features in densely forested mountains by very high-resolution airborne LiDAR survey. Geomorphology 182:104–115
13. Meigs A (2013) Active tectonics and the LiDAR revolution. Lithosphere 5:226–229
14. Wagenbrenner J, Robichaud P (2014) Post-fire bedload sediment delivery across spatial scales in the interior western United States. Earth Surf Proc Land 39:865–876
15. Westoby MJ, Brasington J, Glasser NF, Hambrey MJ, Reynolds JM (2012) 'Structure-from-Motion' photogrammetry: a low-cost, effective tool for geoscience applications. Geomorphology 179:300–314.
16. Young AP, Olsen MJ, Driscoll N, Flick RE, Gutierrez R, Guza RT, Johnstone E, Kuester F (2010) Comparison of airborne and terrestrial lidar estimates of Seacliff erosion in Southern California. Photogramm Eng Remote Sens 4:421–427

Abstract

The second chapter presents the generation process of pointcloud in geomorphology. The chapter starts with laser technologies that have given rise to aerial laser scanner (ALS) and terrestrial laser scanners (TLS) and how they have emerged in geomorphology, notably through the needs in fluvial geomorphology to detect minute topographic changes, and so over long channel stretches. The characteristics of lasers and the different types of laser solutions are presented with the RIEGL sensors and one of the Leica sensors. Then follows a more practical section explaining how to collect your pointcloud and how to generate it. This first section on laser technologies is then followed by a section on SfM-MVS (structure-from-motion–multiple-view-stereophotogrammetry), with a section on the emergence of SfM-MVS in geosciences and a practical section on how to collect data in the field and the processing steps in Metashape® to generate the pointcloud data. The chapter ends with "photogrammology", which is a non-quantitative use of SfM-MVS in order to retrieve shapes or stitch images, when otherwise necessary information is not available. The chapter includes worked examples.

Learning Outcomes: History, Laser Scanners and Photogrammetry

- Have a basic understanding of the history and evolution of point-cloud technologies applied in geomorphology;
- Know the laser-based (ALS and TLS) and the photogrammetric SfM-MVS point-cloud generation methods;
- Understand the limitations and advantages of the different methods and when using one over another;
- Know the different types of laser scanners available;
- Be able to take photographs for SfM-MVS and process the data into a point-cloud;
- Know what and how to take tie points for laser data and SfM-MVS data as well as ground control points to calculate the model errors;
- Be able to use SfM-MVS for qualitative geomorphology purposes;

Introduction

In this chapter, you will discover two methods to create point-clouds for geomorphological purposes, using light distance and ranging (LiDAR) and structure from motion–multiple-view photogrammetry (SfM-MVS). There are other methods available to scientists, practitioners and students, and students may wonder why choosing SfM-MVS out of all the photogrammetric methods. First of all, constructing this book as a manual, it was necessary to make choices. Secondly, I have chosen techniques that are rapidly growing and are relatively easy to access, so that all your energy is not spent on the tools and the methods, but on geomorphology itself. More specifically, SfM-MVS being a method that allows geomorphologists who are not specialists of photogrammetry to obtain high-quality point-clouds providing that they follow a set of steps (cf. the acquisition and processing protocol proposed by James et al. [16]), and so with virtually no or very little funding.

Chapter structure The present chapter starts with the different types of LiDAR technologies in geomorphology and when you should use one over another. Then, it presents SfM-MVS as a photogrammetric method for geomorphologists, presenting a protocol that has been suggested and also the different steps and important points to remember to successfully create a pointcloud. At the end of the chapter, I provide some acknowledgement to the fact that geomorphology is also a qualitative science (this is not a pejorative word), and I introduce a discussion of how SfM-MVS can be used as a "photogrammology" to "recognize shapes".

© Springer Nature Switzerland AG 2022

C. Gomez, *Point Cloud Technologies for Geomorphologists*, Springer Textbooks in Earth Sciences, Geography and Environment, https://doi.org/10.1007/978-3-031-10977-5_2

Who is this chapter for? This chapter is tailored for undergraduate students and graduate students of geoscience and geography, as well as more advanced scientists and practitioners who are not used to collecting data to construct point-clouds of surfaces. It provides a description of the tools available (it is not an exhaustive list of course), before presenting some the "how-to", so that the reader gets a good hand-on experience. But, before starting with those technical and methodological aspect, we start with a rapid introduction of point-cloud technology in geomorphology, with a few examples in different environments, acknowledging that those will be developed in the chapters dedicated to hillslope geomorphology, coastal geomorphology, volcanic geomorphology and river and floodplain geomorphology.

Who is this chapter not for? If you are already a user of LiDAR technologies and SfM-MVS, and that you already have a good command of the data acquisition process, the present chapter may just be a simple reminder for you, and you may want to skim it. You can then go to the next chapter on point-cloud processing. For those of you who are familiar with laser scanning and point-cloud acquisition with this technique but would like to obtain a deeper understanding of its usage and applications in geosciences and environmental sciences, I would like to refer you to the excellent book edited by Heritage and Large [15].

2.1 LiDAR Point-Cloud in Geomorphology

2.1.1 The Introduction of Laser Technologies in Geomorphology

Although not the cheapest option, light detection and ranging (LiDAR) is certainly the best existing method to collect point-clouds in geomorphology, not only because it does not suffer number of the distortions that occur with the photogrammetric method SfM-MVS, but also because it can "see through" the vegetation and in particular cases through shallow water. Laser scanner is therefore particularly indicated for geomorphological purposes.

In geomorphology, LiDAR has begun, not as much as the result of technological development within the community of geomorphologists itself, but as the result of the integration of technologies that had arisen in other scientific and engineering fields. And, this external influence that arose from cross-disciplinary technologies has grown beyond the simple modification of the geomorphologist's toolbox. It has—like structure from motion at the time I am writing those lines—contributed to, and reshaped the development of geomorphology and the questions it can ask as a science. Although I never felt comfortable with the idea that tools are shaping a

scientific field, one has to recognize that point-cloud technology has just done that. For instance, the counterslopes just above the settlement of Franz Joseph on the West Coast of New Zealand, which are a sign of large-scale slope movements, could have never been investigated without the help of LiDAR technology. The technology has allowed the extraction of the topography from the thick vegetation on extremely hard-to-access steep slopes, revealing otherwise invisible man-deep scarps at the top of the slopes. Thus, we can then only agree that this technological input in the field of geomorphology has had the effect of a real revolution, with the possibility to start new types of research, and the opportunity to measure landscape features at scales and locations not accessible otherwise.

Indeed, when I was doing my classes of geomorphology at Sorbonne University in Paris, in the 1990s and early 2000s, we were still taught that obtaining topographic data over large areas was synonym of a trade-off with measurement density, whereas micro-topographic and site-scale data could be done with greater details with more measurements. What LiDAR has done is to break this scale relation. The density of data that you can use is now more limited by the capacity of the computing station that you are using, rather than the field limitations (although they still exist). For instance, in 2020, the USA is on track of having a country-wide cover of LiDAR data, freely available within the country for research purposes. In the UK, such a database already exists, with data spread over several years, even allowing chronological analysis. In other words, geomorphology has also entered the age of big data, where the paradigm has shifted from data acquisition to data assimilation, processing and interpretation. For many years, geomorphologists have been dreaming about what could be done with a more complete dataset in one or another location. Today, we often have it, and eventually freely. The question is now how can we use it, what can we state and demonstrate from these datasets? What questions can we ask that we could not ask before?

The very beginning of LiDAR technology certainly did not spring from one single laboratory, but if I was to just name one very influential group, NASA has played a significant and instrumental role in LiDAR development. It started with Arctic surveys in the 1960s, when the technology was then a military-grade tool, which had been developed to estimate the distance of enemies and calibrate fire. Its use spilled over from military purposes to others from mid-1970s and onward. This was an era of "single shots" and single measurement LiDAR. When it comes to scanners, we have to wait for the end of the twentieth century, when laser technology for terrain mapping emerged: first in Germany at the Surveying and Mapping Agency (SMA) of the Federal States of Germany. The German researchers collected multiple laser returns from an airborne laser scanner and developed methods to differentiate buildings from veg-etation and the ground, generating what was a promising new type of datasets. Some of these datasets were collected over the regions of Bavaria and North-Rhine-Westphalia, with pulse rates at 2000 Hz, scan rates between 7 and 12 Hz, scan angles between 9 and 20° at height of 900 m above the ground. These campaigns occurred during the period of 1994–1999 [36], and it would open the gates to a new era in land surface data acquisition, which is still having reper-cussions in 2020, and is still the object of new development and research. This German technological advance rapidly crossed the channel to become a major research direction of geomorphology in the UK. Indeed, British researchers star-ted to experiment with laser scanners, and a group that originally started at Cambridge University has brought numbers of novel research in geomorphology using this technology, notably fluvial geomorphology.

Since then, point-cloud technology adaptation and application in geomorphology have "exploded" and LiDAR-derived point-clouds have found their way to most of the sub-fields of geomorphology, being now combined with other types of data. For instance, in the sub-field of coastal geomorphology, along the Mediterranean coast of La Mata-Guardamar in the Alicante region of Spain, Pagan et al. [31] recorded 154 ground control points (GPCPs) with a Global Navigation Satellite System (GNSS) in 2017 and compared the data with LiDAR from 2016 and other DEM data dating back to 2001. Using those multiple tools and methods, the authors have notably shown that the overall sand loss was 143,561 m^3 for the period 2001–2017 on the dune field. This data was generated from a large footprint but low vertical variability location, for which high accuracy and precision is essential. Taking another coastal example, the coastal cliffs of the Marsden Bay in the UK have been measured with a combination of terrestrial laser scanner and SfM-MVS calibrated with GCPs at the top and bottom of the cliffs horizontally spaced by a maximum of the cliff height. This has been used to detect rockfall-related evolution of the cliff face with a minimum erosion threshold as small as 0.07 m^3 [48]. From this example, it is then easy to under-stand that the limited amount of information contained in traditional 2.5D DEMs and topographic maps can be blown in a new breadth of information by adding 3D information from the sub-vertical cliff faces. You cannot climb the cliff with a GNSS to take points, and using a total station to image, the surface would require a lifetime of work, during which the cliff face would have changed anyway. Using other photogrammetric methods requires you to know the position of several reference points on the cliff, which is again impractical and most certainly not in line with health and safety regulations. This coastal cliff is one example of why LiDAR point-cloud has been nothing short of a revo-lution in geomorphology.

This idea was relayed by Johnson et al. [18], who called the rise of LiDAR a revolution in "Unravelling Scandinavian geomorphology: The LiDAR revolution". Furthermore, they started their contribution with "In the observational sciences, technical advances are often followed by dramatic increases in scientific discoveries and improved theories […]" (p. 245), which is arguably what has occurred with point-clouds and geomorphology.

2.1.2 LiDAR—How Does It Work?

A LiDAR system is invariably made of (1) the laser pulse system, which physically excites a material such as Gallium Arsenide to generate the energy, which is then pumped into the laser generator (cf. [35] for a more detailed introduction), (2) the optical mechanics that rotate a mirror, in which the laser reflects and spread in a swath, (3) the telescope optical lens, through which the laser is concentrated and propagated towards the target, (4) the optics and the filter to receive the return signal and (5) the sensor connected to the electronic system that limits the noise and convert the received pulses into computational data.

Basic concept—measuring distances The laser and the optical system can either generate simple collimated laser or more complex modulated beams (for instance, the "green LiDAR" that can go through shallow waters, or multi-spectral LiDAR that can also convey information on the nature of the material investigated). The measuring vector is therefore an electromagnetic radiation with a given energy ($E[ev]$) that vary depending on the wavelength ($\delta[m]$), the velocity of the electromagnetic radiation (v) and the Planck's constant (h), so that:

$$E = \frac{h*v}{\delta} \qquad (2.1)$$

The emitted laser can either be a discrete laser, from which the time of flight (TOF) is measured, or it can be a continuous laser, from which phase interferometry for dif-ferent frequencies is calculated. However, numerous topo-graphic laser applications are based on TOF. TOF is based on the precise measurement of the time it takes for the laser to travel the distance between the instrument and the target and back, so that the range (R) is calculated from the velocity of the laser in the air v, and half the time (t) it took to reach and return from the target:

$$R = \frac{1}{2}v*t \qquad (2.2)$$

If the pulsed laser systems use the discrete TOF of each pulse, the phase comparison method is based on the emis-sion of a continuous beam. The continuous beam is also reflected by the object it encounters, but it returns as a series of waves. The signal is modulated over different frequencies, and the different phase differences for different frequencies are resolved as a set of equations (namely the ambiguity resolution) to find the range.

As LiDAR technology is based on the measurement of multiple points from a fixed or a moving emitting station, it translates into a coordinate system with the point of origin being the laser scanner, and the horizontal distance D and the vertical distance, H being simply formulated from the trigonometric angle relations using the range R and the values of the mirror angle, including a vertical and a horizontal angle (Eqs. 2.4, 2.5 and 2.6). The height of the laser scanner (when using a terrestrial laser scanner) is then compensated. In the case of an airborne platform, the movement of the mirror or the prism is used to create series of tracks that are perpendicular to the flight path and then added as successions to the data of flight direction, speed and altitude, from which a series of terrain profiles is generated (actually if you zoom on a dataset generated from an airborne platform, for instance, you can follow the imaginary lines created transversally to the plane path).

The laser footprint size and return intensity In theory, the laser footprint on an object perpendicular to the laser beam should be the size of the collimated laser beam of diameter d_{clb}, but in reality the laser will diverge by a diversion angle ($\theta[\text{rad}]$) which will vary with the range R, and eventually the atmospheric noise (N_{atm}) for long-range lasers, so that the footprint surface Fs is:

$$\text{Fs} = \frac{\prod(\theta R + d_{\text{clb}} + N_{\text{atm}})^2}{2} \qquad (2.3)$$

The spread of the footprint is further complicated by the surface it reflects on. This one can be at different angles from the beam and the rugosity of the surface. Combined with the range, it influences the intensity of the reflection, so that different materials will reflect different proportions of the emitted signal. This characteristic will be very important for signal processing, as it will allow us to divide the point-cloud generated between for instance roof material, vegetation and road surfaces for instance.

2.1.3   The Different Types of LiDAR

In this section, I invite you to differentiate airborne platforms from ground platforms, as this difference will certainly be a main concern for the geomorphologist. This dichotomy is not the sole way to differentiate LiDAR technology. Laser specialists may certainly choose other differentiation, notably based on the type of laser and its energy, but for geomorphological purposes, it is an adequate division.

Before starting this platform-based differentiation, the reader should note that laser systems should be used with caution depending on their class. If you are using laser technology, i.e. a beam of energy, there are safety concerns involved, and lasers have been divided into different classes, notably as they can cause damage to human eyes and skin (Table 2.1).

Airborne LiDAR: Airborne Laser Scanner (ALS) Airborne laser scanner (or ALS) is a tool that captures a series of laser returns from a swath using a back and forth mirror movement, perpendicular to the flight direction. Recording the flights' parameters from the plane, the swaths are stitched together along the flight path creating strips of scanned point-clouds. Geomorphologists usually don't have to deal with data acquisition and construction, which are performed by specialized companies or national organizations. In the Treatise on Geomorphology (2013), Schrott et al. wrote in their chapter on LiDAR: "As ALS data acquisition is generally done by private companies or public authorities, this section—dealing with practical issues regarding data acquisition and processing—focuses on TLS systems [...]". In other words, geomorphologists will usually receive either a DEM/DSM or .LAS and .LAZ format data to work from, and this chapter assumes that it is the case more often than none. Readers who are interested in the processing of raw data and construction of the LAS and LAZ dataset, please refer to existing engineering and engineering surveying manuals.

ALS is the evolution of the airborne laser profilers (either on board planes or on artificial satellites). These systems have evolved into scanning systems, notably thanks to the integration of rotating mirrors that can propagate the laser beam in swaths.

ALS leverages the movement of the aircraft/spacecraft to collect data over large areas. ALS is therefore an attractive method when the scientist is confronted with large areas to map, as LiDAR mounted on an aircraft provides acquisition rate that can be up to 90 km^2/h [28]. Because ordering an ALS flight can be costly, cost–benefit analysis only becomes advantageous when several square kilometres need to be captured. When working on steep slopes or on volcanoes, ALS is often used from helicopters instead of traditional planes, because it allows to reduce the velocity and concentrate data collection in areas of interests. At Unzen Volcano in Japan for instance, the LiDAR data ordered by the government has been in part surveyed from helicopters.

ALS comprises the laser scanner itself (cf. Table 2.2 for some of the products proposed by RIEGL for instance), the GNSS antenna connected to the inertial measurement unit (IMU), and a data processing and recording device. The GNSS antenna on the plane is also complemented with a GNSS antenna (or several) on the ground, in order to correct

Table 2.1 Simplified description of four laser classes (there are other subdivisions) and the main differences associated with each class

Laser class	Simplified description
I	Lasers that do not emit radiation and are not hazardous
II	Visible lasers of radiant power < 1 mW
IIIA	Lasers that are < 5 mW and that can be hazardous to view, but require limited control
IIIB	Lasers of power between 5 and 500 mW (quite a broad range). Those can be dangerous for the eyes and operation should be done according to regulation
IV	These lasers are powerful enough to start a fire, burn skin and eyes permanently and strict safety measures need to be obeyed

A long-range TLS requires a higher level of energy, and thus, it is being more dangerous to human eyes and skin. Therefore, you always want to operate strictly within the rules of health and safety and the regulations of the place you are using the TLS. The choice of TLS will depend on the object to be studied

the moving airborne GNSS antenna for error using real-time kinematic (RTK) or postprocessing kinematic (PPK) methods. The IMU combined with the GNSS data then allows correction for the position of the aircraft, the velocity, the roll and pitch (there is never a truly straight flight) and the true heading. It is therefore essential that the clock of the different instruments is perfectly synchronized for all the measurements to overlap at the same time. This seemingly trivial task is actually complicated, especially with different instruments working at different frequencies, and transmitting data at different rates, etc. For a list of the different providers and existing systems and their customization, please see Chap. 2 in [43].

ALS is a very attractive topographic solution for geomorphologists—except for the cost—but it is not a silver bullet, especially in wet and often rainy climate, as the laser may fail to penetrate the cloud cover. Furthermore, rainfall and atmospheric moisture generate laser beam scattering. Winds and pressure differences during a storm, a rainy event or when the air rises at the end of the afternoon in subtropical climates, all have an impact on the aircraft stability, and the data collected. ALS from traditional aircraft is best suited when the footprint to record is several square kilometres, but for smaller areas, UAV-based LiDAR and TLS are certainly a better adapted option, especially in complex terrain.

Terrestrial LiDAR: Terrestrial Laser Scanner (TLS)
Building on the success of ALS, TLS is the latest add-on to the geomorphologist's toolbox. It is a smaller scale equivalent to the ALS, and instead of being mounted on a plane, it is often mounted on a tripod and equipped with a vertical and a horizontal rotating screw, providing up to 360 scanning options. TLS was first developed in 1999, and with just a little more than twenty years under the hood, there are still many technical aspects that need standardizing and testing, especially because the instruments keep evolving and changing at a rapid pace. Only 6 years ago (in 2014), Kociuba et al. [19] were writing that "the application of TLS in studies of the cold-climate environments is still at the

stage of testing". Since then, the evolution has been relatively slow, because the international economy slowing down in most regions of the world has meant that fewer funding has been available at research institutions and at national level. In Japan, for instance, and despite the common image overseas, buying a TLS is beyond the scope of most of the national grants available and the average university funding. Funding at university level has been cut by 1% every year for more than a decade, leaving only a handful of flagship institutions with the ability to obtain the equipment. With the spread of COVID-19 in the early 2020s, such conditions have become the norm in numerous countries and the pace of progress has dropped dramatically, even in countries that used to lead the charge.

TLS can be either static, mounted on a tripod, or it can be mobile when mounted on a vehicle in combination with an IMU. It is the first option that is often used in geomorphology, with several scans performed from a set of static locations. The scanning methodology is indeed based on a series of overlapping measurements from different vantage positions. The scans are meshed together using a minimum of three targets that can be recognized in the different point-clouds (usually spherical balls mounted on poles). Please note that the distortions are minimal and three targets are often sufficient, but when we will look at structure from motion, it will be necessary to acquire more targets and place them in "strategic" locations to reduce the error.

Most scanners do not offer the full 360 dome-like reconstruction. Those offering a full 360° view are named panorama scanners, while the systems that only scan in one direction are named camera scanners. In between these two end-members, there is a variety of systems that scan the surrounding between given angles (typically the foot of the laser will not be scanned, for physical reasons). For example, the RIEGL VZ-400i makes measurements up to a range of 800 m and for a field of view of 100° by an azimuthal 360° (for a 5 mm precision and an accuracy of 5 mm). Thus, for each point measurement, it includes two angles and the range, and depending on the options, the intensity of return

Table. 2.2 Different types of LiDAR systems developed by RIEGL to capture point-clouds, airborne, on the ground fixed or on cars and other vehicles (photographs and data courtesy of RIEGL)

	Product	Key features	Picture
TLS	*RIEGL* VZ-400i	*Ultra-high-performance 3D laser scanner* • Innovative processing architecture for data acquisition and simultaneous geo-referencing in real-time • High laser pulse repetition rate of up to 1.2 MHz • High-speed data acquisition with up to 500,000 meas./s • Wide field of view 100° × 360° • Range up to 800 m • Accuracy 5 mm, precision 3 mm • Automatic on-board registration • Optional integration of camera • Main dimensions: 206 mm × 346 mm (width × height) • Weight: approximately 9.7 kg (with antennas)	
TLS	*RIEGL* VZ-2000i	*Long range, very high-speed 3D laser scanner* • Innovative processing architecture for data acquisition and simultaneous geo-referencing in real-time • High laser pulse repetition rate of up to 1.2 MHz • High-speed data acquisition up to 500,000 meas./s • Wide field of view, 100° × 360° • Range up to 2500 m • Accuracy 5 mm, precision 3 mm • Automatic on-board registration • Optional integration of camera • Main dimensions: 206 mm × 346 mm (width × height) • Weight: approximately 9.8 kg (with antennas)	
TLS	*RIEGL* VZ-6000/*RIEGL* VZ-4000	*3D ultra-long-range terrestrial laser scanners* • *RIEGL* VZ-6000: more than 6000 m measurement range • *RIEGL* VZ-4000: very long range up to 4000 m • Wide field of view, 60° × 360° • High-speed data acquisition up to 222,000 meas./s • Accuracy 15 mm, precision 10 mm • Built-in calibrated 5-megapixel camera • Exceptionally well suited for measuring snowy and icy terrain • Main dimensions (L × W × H): 248 × 226 × 450 mm • Weight: approximately 14.5 kg	
ALS	*RIEGL* VUX-1LR	*Lightweight airborne laser scanner* • 15 mm survey-grade accuracy • Scan speed up to 200 scans/s • Measurement rate up to 750,000 meas./s • Operating flight altitude more than 1700 ft • Field of view up to 330° for practically unrestricted data acquisition • Accuracy 15 mm, precision 10 mm • Main dimensions VUX-1UAV without/with cooling fan: 227 × 180 × 125 mm/227 × 209 × 129 mm • Weight VUX-1UAV without/with cooling fan: approximately 3.5 kg/approximately 3.75 kg	

Table. 2.2 (continued)

	Product	Key features	Picture
ALS	*RIEGL* VQ-780 II	*Waveform processing airborne laser scanner for wide area mapping and high productivity* • High laser pulse repetition rate up to 2 MHz • Up to 1.33 million meas./s on the ground • Multiple-time-around (MTA) processing of up to 35 pulses simultaneously in the air • Accuracy 20 mm, precision 20 mm • Interface for GNSS time synchronization • Main dimensions (L × W × H): 425 mm × 212 mm × 331 mm • Weight: approximately 20 kg	
ALS	*RIEGL* VQ-480 II/*RIEGL* VQ-580 II	*Waveform processing airborne laser scanning system* • High laser pulse repetition rate, up to 2 MHz • Measurement rate up to 1,250,000 meas./s • Compact and lightweight design also ready for integration in UAVs with higher payload capacity • Wide field of view of 75° • Accuracy 20 mm, precision 20 mm • Interfaces for up to five optional cameras • *RIEGL* VQ-480 II works at 1550 nm wavelength *RIEGL* VQ-580 II works at 1064 nm wavelength • Main dimensions: 378 mm × 193 mm × 252 mm (L × W × H, without mounted carrying handles) • Weight without integrated IMU/GNSS: 10.1 kg With integrated IMU/GNSS: 10.5 kg	
ALS	*RIEGL* VQ-1560 II/*RIEGL* VQ-1560 II-S	*Dual channel waveform processing airborne LiDAR scanning system for high-point density mapping and ultra-wide area mapping* • High laser pulse repetition rate up to 4 MHz • Up to 2.66 million meas./s on the ground • *RIEGL* VQ-1560 II: enables multiple-time-around (MTA) processing of up to 35 pulses simultaneously in the air • *RIEGL* VQ-1560 II-S: enables multiple-time-around (MTA) processing of up to 45 pulses simultaneously in the air • Accuracy 20 mm, precision 20 mm • Integrated inertial measurement unit and GNSS receiver • Integrated, easily accessible medium format camera • Main dimensions: Ø 524 mm × 780 mm (flange diameter × height, without flange mounted carrying handles) • Weight: approximately 55 kg without any camera but including a typical IMU/GNSS unit Approximately 60 kg with optional components	

(continued)

Table. 2.2 (continued)

	Product	Key features	Picture
ALS	*RIEGL* VQ-1560i-DW	*Dual wavelength waveform processing airborne LiDAR scanning system for high-point density mapping applications* • Enhanced target characterization based upon simultaneous measurements at green and infrared laser wavelengths. These wavelengths are well chosen to allow the acquisition of scan data of complementary information content, thus delivering two independent reflectance distribution maps, one per laser wavelength • Integrated inertial measurement unit and GNSS receiver • Integrated, easily accessible medium format camera • Main dimensions: Ø 524 mm × 780 mm (flange diameter × height, without flange mounted carrying handles) • Weight: approximately 60 kg without any camera but including a typical IMU/GNSS unit Approximately 65 kg with optional components	
ALS	*RIEGL* VQ-880-G II/*RIEGL* VQ-880-GH	*Topo-bathymetric airborne laser scanning system with online waveform processing and full waveform recording* • Designed for combined topographic and hydrographic airborne survey • Green laser channel with up to 700 kHz measurement rate • Laser channel with up to 279 kHz measurement rate and improved ranging performance • Accuracy 25 mm, precision 25 mm • Integrated inertial navigation system • Up to two integrated digital cameras • Main dimensions: 489.5 mm × 660 mm × 580 mm, mounting flange 580 mm × 580 mm (flange diameter × height) • Weight: approximately 70 kg (with IMU/GNSS/cameras and optional infrared laser scanner) • VQ-880-GH: form factor with reduced height optimized for helicopter integrations	
ALS/ULS	*RIEGL* VUX-240	*Lightweight UAV laser scanner with online waveform processing* • Laser pulse repetition rate up to 1.8 MHz • Measurement rate up to 1,500,000 meas./s • Scan speed up to 400 lines/s • Operating flight altitude up to 1400 m/4600 ft • Field of view up to 75° • Accuracy 20 mm, precision 15 mm • Interfaces for up to 4 optional cameras • Main dimensions: 292 mm × 164 mm × 185 mm (L × W × H, without IMU/GNSS) 380 mm × 164 mm × 185 mm (with IMU/GNSS) • Weight: ≤ 4.1 kg (without IMU/GNSS), ≤ 4.9 kg (with IMU/GNSS)	
ALS/ULS	*RIEGL* VQ-840-G	*Compact Topo-Bathymetric airborne laser scanner with online waveform processing and full waveform recording* • Designed for combined topographic and bathymetric airborne and UAV-based surveying • High spatial resolution due to measurement rate of up to 200 kHz and high scanning speed of up to 100 scans/s • Accuracy 15 mm, precision 10 mm • Integrated inertial navigation system (optional)	

(continued)

Table. 2.2 (continued)

	Product	Key features	Picture
		• Additional, fully integrated infrared laser rangefinder or integrated digital camera (optional) • Main dimensions: 360 mm × 285 mm × 200 mm (LxWxH) • Weight: approximately 12 kg	
ULS	*RIEGL* miniVUX-1UAV/ *RIEGL* miniVUX-2UAV/ *RIEGL* miniVUX-3UAV	*Extremely lightweight LiDAR sensor for unmanned laser scanning* • *RIEGL* miniVUX-1UAV: measurement rate up to 100,000 meas./s, 100 kHz PRR • *RIEGL* miniVUX-2UAV: measurement rate up to 200,000 meas./s, 100/200 kHz PRR selectable • *RIEGL* miniVUX-3UAV: measurement rate up to 200,000 meas./s, 100/200/300 kHz Laser PRR selectable • Scan speed up to 100 scans/s • 360° field of view • Multiple target capability—up to 5 target echoes per laser shot • Accuracy 15 mm, precision 10 mm • Exceptionally well suited to measure in snowy and icy terrains • Main dimensions (L × W × H)/weight: With cooling fan: 243 × 111 × 85 mm/approximately 1.6 kg Without cooling fan: 243 × 99 × 85 mm/approximately 1.55 kg	
ULS	*RIEGL* VUX-120	*"Nadir/forward/Backward" (NFB) scanning UAV LiDAR sensor, especially suited for high-point density corridor mapping* • Laser pulse repetition rate up to 1.8 MHz • Measurement rate up to 1,500,000 meas./s • Scan speed up to 400 lines/s • Operating flight altitude up to 720 m/2350 ft • Field of view up to 100° • Accuracy 10 mm, precision 5 mm • Nadir/forward/backward scanning for unrivalled completeness of scan data even on vertical structures and narrow canyons • Interfaces for up to 2 external cameras • Main dimensions (L × W × H): 225 mm × 117 mm × 126 mm (without connection box) 242 mm × 117 mm × 126 mm (with connection box) • weight (without Interface Box/with Interface Box): approximately 2 kg/approximately 2.2 kg	
ULS	*RIEGL* VUX-1UAV	*Versatile UAV laser scanner with online waveform processing* • Scan speed up to 200 scans/s • Measurement rate up to 500,000 meas./s (@ 550 kHz PRR and 330° FOV) • Operating flight altitude more than 1000 ft • Field of view up to 330° for practically unrestricted data acquisition • Accuracy 10 mm, precision 5 mm • Main dimensions: VUX-1UAV without/with cooling fan 227 × 180 × 125 mm/227 × 209 × 129 mm • Weight: VUX-1UAV without/with cooling fan Approximately 3.5 kg/approximately 3.75 kg	

(continued)

Table. 2.2 (continued)

	Product	Key features	Picture
ULS	RiCOPTER with VUX-SYS	*Remotely piloted aircraft system for unmanned laser scanning* • RiCOPTER remotely piloted aircraft system equipped with *RIEGL* VUX-SYS complete miniaturized, lightweight ALS System • *RIEGL* VUX-1UAV lightweight airborne laser scanner fully integrated, providing 230° FOV, an effective measurement rate up to 350,000 meas./s, and 10 mm accuracy • Operates up to 3 digital cameras	
MLS	*RIEGL VUX-1HA*	*High-performance LiDAR sensor for kinematic laser scanning* • Scan speed up to 250 scans/s • Measurement rate up to 1,000,000 meas./s • "Full circle" field of view of 360° for unrestricted data acquisition • Accuracy 5 mm, precision 3 mm • Easily mountable to whatsoever type of moving platform • Main dimensions: VUX-1HA without/with cooling fan 227 × 180 × 125 mm/227 × 209 × 129 mm • Weight: VUX-1HA without/with cooling fan Approximately 3.5 kg/approximately 3.75 kg	
MLS	*RIEGL VMX-2HA*	*High-speed, high-performance dual-scanner mobile mapping system* • High laser pulse repetition rate of up to 2 MHz • 500 scan lines per second • Range 420 m • Field of view 360° • Accuracy 5 mm, precision 3 mm • Camera interface for up to 9 optional cameras	
MLS	*RIEGL VMQ-1HA*	*High-speed single-scanner mobile mapping system* • High laser pulse repetition rate of up to 1 MHz • 250 scan lines per second • Range 420 m • Field of view 360° • Accuracy 5 mm, precision 3 mm • Optional integration of up to 4 cameras	

and the combination with RGB and other values extracted from a camera that can be coupled to the system.

The point-cloud generated from TLS systems can be up to 360° in one scan (depending on the scanner). The results are recorded using a horizontal azimuth mount and a vertical rotation of the mirror that propagates the laser along a vertical plane. Combining the two, we obtain a set of points located on a vertical plane, and this vertical plane is rotated into a set of azimuthal angles, so that a 3D scene can be reconstructed from a vertical (θ_v) and a horizontal angle (θ_h) as well as the range (R), which can be automatically (or by the operator) converted into a Cartesian referential X, Y, Z as follows:

$$X = \cos\ \cos(\theta_h) * R \qquad (2.4)$$

$$Y = \sin\ \sin(\theta_h) * R \qquad (2.5)$$

$$Z = \sin\ \sin(\theta_v) * R \qquad (2.6)$$

Please note that the polar coordinates include the position of the TLS, as well as the free path from the TLS to the target. You can thus compute empty space as well as the locations with objects [14]. Once you turn the dataset into a Cartesian system, even if the X, Y, Z coordinates still include the position of the TLS (potentially at $X = 0$, $Y = 0$, and $Z = 0$), computing the free pathway would require you to compute back the vector, and know which point was detected from which TLS position. This is not a processing step that is commonly done yet, but it is certainly a research direction that the student reading this manual may be interested in developing to create a new algorithm detecting surfaces and targets.

The type of TLS can be further classified depending on its range, with short-, mid- and long-range TLS. Using the line-up of TLS from RIEGL as an example, the RIEGL VZ-400i has an optimal 800 m range, the RIEGL VZ-2000i has a range of 2.5 km and the RIEGL VZ-4000 and VZ-6000 have, respectively, ranges of 4 and 6 km. These different ranges come with a trade-off as the short-range instrument has an accuracy of 5 mm and the long-range systems 15 mm. Moreover, the acquisition speed varies from 500,000 points per second at short range against 220,000 points at long range. Finally, weight is also an important factor for the geomorphologist. If you want to carry the instrument upslope or in a gully, the RIEGL VZ-400i is less than 10 kg while the long-range system comes close to 15 kg. If you add to this weight, the tripod and the other accessories, it is then important to best define your needs before carrying one tool over another to the field. Although the RIEGL VZ-400i already has a maximum range of 800 m, typical short-range laser scanners are rather < 250 m range. For instance, Leica's HDS4500 has a $\sim$ 50 m range, the HDS6000 a 79 m range, the ScanStation and HDS3000 a 300 m range. The Faro LS880 and the Z-F Images5006 also offer ranges < 80 m. If you are interested in purchasing a laser scanner, short-range systems will usually come at a lower price than the longer-range ones. Choosing a TLS will therefore be a compromise between the precision and accuracy you need, the optimal distance necessary for the type of fieldwork you do, the weight you are ready to carry (upslope for instance) and finally the price you can afford to buy the equipment.

2.1.4 Hybrid Solutions Combining Total Station and Laser Scanning

In recent years, "cross-over" systems have also been developed. The Leica MS60 (Fig. 2.1) proposes a hybrid system between a laser scanner and a total station. A good description of this tool would be a full-spec total station with scanning capability. The advantage of a total station is the ability to "choose" the reflectors you want to image and record them as specific points. For instance, if there are key points you want to image, or if in a constantly moving and changing valley (erosion, deposition-intensive proglacial valley for instance), you may want to find a set of points that are relatively stable, on bedrock for instance, to reference your dataset, especially if you are in a narrow valley with limited sky, so that even GNSS is not so helpful.

2.1.5 From Fieldwork to Point-Cloud

In this short section, I am providing you with a step-by-step description of gathering data, from planning to point-cloud generation. The analysis of the point-cloud comes in the next chapter.

Fieldwork: Now that we have been reviewing the different types of laser systems and acknowledging that you will certainly be taking the ALS data yourself, what do you need to do for your TLS fieldwork?

(1) **Planning**: This first step is the most crucial step. At no point, you want to put the TLS on the back of your truck and drive to the field before careful planning. This means that you need to first define the size of the field you want to scan, then decide how many scans you will need, and from which positions you will need to take the scan. It is most likely that you will already have photographs of the area and a crude topographic map. You will need to consider the topography during planning (can you see the environment you want to scan from the chosen scanning locations). Are there trees, buildings, etc., that you will need to move your TLS around? Once you have a crude map of how you are going to scan the area, you also will need to know where you can place the tie points (reflective balls on a stick) and how many reflectors you will need (in forest environments, reflectors might become essential), and where to place them. Then, consider the day and time of the day you are doing the scans, will there be a lot of people around, will you need to cordon off the area if your laser is not eye-safe, and what are the authorizations and paperwork you need to do beforehand (those always take more time than you think, so plan well in advance). Once you have a plan, do not forget the health and safety aspect of your trip and make sure that you follow the regulations about using the laser scanner. Using all this information, you will then be able to know how much time it will take you to generate the different scans.

(2) **Getting into the field**: Prepare your equipment, charge the batteries and make a checklist of the equipment you will need in the field, going through each of the tasks one by one. It is always a good idea to have spare targets, batteries, etc., but also make sure to bring a notebook with you and a small camera (a smartphone usually does it) to document your process. One thing I do in the field is to take a photograph of the notebook where I explain what is being done before taking a photograph of the scene. As a rule of thumb, getting into the field, you will need (a) your TLS, (b) the tripod to mount the TLS, (c) a

Fig. 2.1 The Leica MS60 in the field in New Zealand (South Island) near Porter skifield (**a**) and lake Lyndon in the upper country of Canterbury. Although the laser scanning capability has a shorter range, (**b**) the total station could record control points 650 m away on the wall of a cottage (the reader will note that this performance is subject to illumination, moisture, pollution and other factors that can affect the measurement)

ruler to measure the height of the mount and tripod if it is not automated, (d) a set of targets reflectors and a set of balls mounted on tripods or sticks, (e) batteries and the battery charger for your vehicle or place you will stay, (f) memory sticks and cabling to transfer the data if you plan to stay overnight, as well as the laptop to get the data onto; (g) a set of covers against the weather (plastic trash bags usually do the trick); (h) a hand-held GPS; (i) a notebook and your smartphone and camera. Then, you will have environment-specific elements. If you work on hard-rock surfaces, you may need to hammer or drill an anchor for your tripod (to avoid it moving in the wind, etc.), and you may also need blunt instruments to cut vegetation and tree branches, etc.

(3) **In the field**: Providing that you prepared your equipment successfully and that everything is ready for you to be brought in the field, make sure that you follow your plan and that you make any note of the changes. You will then install your TLS at one or a set of locations and run the TLS. If it is the first time you use the TLS in the given environment, you may want to extract one of the scanned data on your laptop and check whether you are collecting the data successfully, or whether there are unanticipated problems. While you are running your scan, take notes of what is being recorded, the exact location of your scanner and take photographs that will help you when you are back at the laboratory. Make sure that you take notes of the start and end time of each scan and that you also record the names of the scans you are doing separately in your notebook. Make sure that you also record the weather, the day, who was with you on the day, so that it is easier to remember things and you can also ask if there is any information you may have missed.

(4) **Cleaning**: Once you have taken your scan, or several of them, make sure that you clean the area you have been working on and compare your equipment list to what you are bringing back, to make sure you do not forget anything. Always put back the instrument in its hard-box before transporting it and make sure that you follow the maker specifications to avoid damaging the equipment. In Japan and Indonesia where I work, humidity is often the enemy, so you may want to check on and add extra-silica gel to maintain the equipment in dry conditions. You are then ready to go back to your laboratory and think about generating the point-cloud.

Generating the point-cloud: Now that you are back at the laboratory (or done with a first set of measurements in the field) with your data in one format or another, it is most likely that you have the results of one or several scans, and there will be a few steps to take before you can turn towards "proper" data processing. We will see those in more detail in Chap. 3, but for the sake of providing you with a "full workflow" here are the main three processes you still need to do (to which error analysis may be added as well).

(5) **Registration Process**: The first step will be the registration process. This step consists in stitching together the different scans you may have taken from different locations, all together in one single point-cloud. When doing the registration process, you should be looking into having an error inferior to a few millimeters. As this error will propagate in all the following processing steps, the lower the error the better, regardless of what you are attempting to do with the dataset later on. Although registration software does not need targets and other artefacts in most cases, the targets and spheres you placed in the field can be used to improve the registration process. You should, however, be ready to spend

some time doing the registration. It is not a difficult task, but it is time-consuming, as in most cases even crude automated alignment will need some manual cross-check. Furthermore, the larger the point-cloud the more computation intensive the process becomes and even with a dedicated computer, it may be a slow process.

(6) **Georeferencing process**: Unless you had a GNSS with PPK or RTK functionalities attached to your laser scanner, it is most likely that the internal GPS has given you a position within a few metres of accuracy. If there are known points in the field you are working, you may want to have a target at this location to then georeferenced the point-cloud from them. These targets can then be used as tie points (to match two point-clouds, to georeference it, etc.) or as ground control points (GCPs), points you will use to measure the potential error and bias in your dataset (see Chap. 3 on these issues).

(7) **Visualization**: Once you have done the two first steps, your point-cloud is now ready for visualization and basic measurements (angles, etc.) from the point-cloud.

2.2   SfM-MVS: Structure from Motion—Multiple-View Stereophotogrammetry

SfM-MVS stands for structure from motion—multiple-views stereophotogrammetry. This acronym is the bundle of two different approaches. SfM reconstructs a 3D scene from 2D images as it computes the geometric characteristics of the camera. MVS, in turn, uses the SfM data to perform a multiple-views photogrammetric calculation of the elements that were not picked up by the SfM process (Fig. 2.2). SfM will usually produce what is often called a "sparse"

point-cloud, while the MVS will have the effect of densifying this point-cloud, by using the camera information and the 3D scene information produced by SfM.

SfM is thus the process of reconstructing the 3D scene and finding the camera location, pitch, roll and orientation. As it is an iterative method, the position of the camera becomes more and more accurate as the algorithm runs over a larger amount of photographs. If you use either the Photoscan Pro software or the Methashape Pro software produced by Agisoft notably, you can also specify the location of the camera, its roll, pitch and orientation. But if this helps constraining the positions of the cameras as it reduces the scale of uncertainties, the SfM algorithm will still attempt to calculate the "best" location in terms of computed geometry. This is the reason why even if you enter camera data, the software will give you data on the error. This is also true for tie points you can specify in a scene. Even if the software will use the tie points to help the process, it will also provide you with a RMSE value, indicating what the calculation produced compared to the data you input.

One could think that SfM would then work without any input (except at a further step for error calculation), but you have to remember that SfM is not scale-dependent and without any input about the camera geometry or a set of given tie points, there is absolutely no way for the software to know whether, let's say, you are trying to reconstruct a toy-mountain of a few tens of centimetres from a one that would be several thousand metres high.

The next question you may then ask yourself is: how does the algorithm match different photographs together, if there aren't any geometric elements tying the 3D scene to the photographs? For this purpose, the SfM algorithm starts with recognizing elements from one photograph to another, and for a same object, because the camera can be very near or very far (resulting in different representations on the

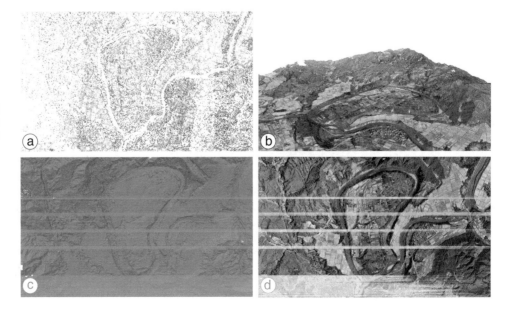

Fig. 2.2 Typical products created by the SfM-MVS workflow (here in the software Agisoft PhotoscanPro—now replaced by Metashape Professional) with **a** the sparse point-cloud generated by SfM, **b** the dense point-cloud generated by MVS, **c** the triangulation of the pointcloud into a mesh (here 2.5D instead of full 3D), and **d** projection of the imagery over the meshed data to generate otherthophotograhs

camera), scale-invariant algorithms have been developed to match from one photograph to another objects that otherwise may not look the same due to the distance, but also the changes in angle, etc. One of the most popular and the first algorithm to do so is the SIFT algorithm [37], which stands for scale-invariant features transforms. From this pioneer, other methods have emerged such as the ASIFT and MOPS algorithms, which are both scale-invariant object detections systems. With the improvement of computing capacities, machine learning has also been employed to generate detectors, with notably the FAST and the AGAST algorithms.

As you may want to try by yourself what is being detected during the SfM process, notably using the SIFT algorithm, and then understand what features may be matched with which one, in the following example, you are going to use MATLAB (but the same library also exists for python and C), with the vlfeat library (https://www.vlfeat.org/overview/sift.html). The online portal is very well documented, and you just have to follow the instructions to install the library. The downloadable package also includes examples and demos, which will guide you through a complete mastery of the package.

First, open MATLAB and navigate using the left pane on the figure underneath (Fig. 2.3) to go to the folder where the "toolbox" folder is. From this folder, just import the library using:

>> run('vl_setup')

You can then open your favourite photograph using the prompt or by just dragging a photograph to the "workspace" pane on the right.

Then, you will need to change the format of the image into a grey scale that can be used for the SIFT algorithm (the photograph I used was named IM_6041):

GscaleImage = single(rgb2gray(Im_6041));

```
[a,b] = vl_sift(GscaleImage);
```

According to the SIFT algorithm tutorial, you can then overlay the SIFT results on top of the photograph, as follows using 100 descriptors (this section of the code is available at the following page by the authors of the vlfeat algorithm: https://www.vlfeat.org/overview/sift.html):

Fig. 2.3 One of the default MATLAB GUI. Please note that it is very similar to the R-studio we are using in other examples or the Spyder environment for Python (image of the GUI in 2020, one will note that this may change over time)

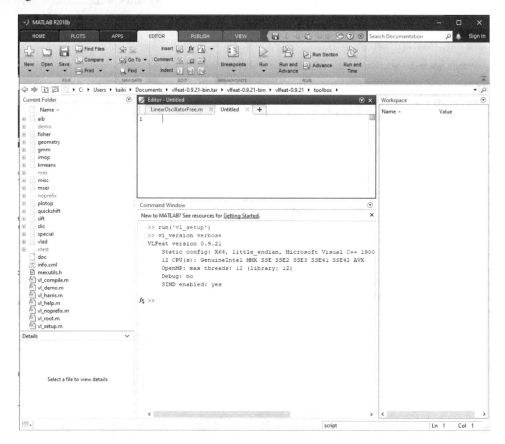

```
Image(GscaleImage)
perm = randperm(size(a, 2));
sel = perm(1:100);
h1 = vl_plotframe(a(:, sel));
h2 = vl_plotframe(a(:, sel));
set(h1,'color','b','linewidth', 3);
set(h2,'color','y','linewidth', 2);


h3 = vl_plotsiftdescriptor(b(:, sel),a(:, sel));
set(h3,'color','g');
```

It results in a set of descriptors for each image that can be shown on the photograph you are analyzing (Fig. 2.4) with corresponding location, scale, and a display of the main gradients.

This set of data can then be used to match sets of photographs, in which case for two photographs, you can run the following lines:

>> I1 = single(rgb2gray(DSC00949));
>> I2 = single(rgb2gray(DSC00950));
>> [a1, b1] = vl_sift(I1);
>> [a2, b2] = vl_sift(I2);
>> [matches, scores] = vl_ubcmatch(b1, b2);

You should not expect to run those steps for your every-day data processing, but it provides you with guidelines on whether the SIFT algorithm can find descriptors in the photographs you are working with. Especially, in geomorphololgy, greyish similar looking gravels for instance, or changing very sharp contrast can be challenges for SfM-MVS, and it is a good way to test whether the algorithms will perform well or not.

Fig. 2.4 Overlay of 100 descriptors over the original photograph, using the vlfeat implementation of the SIFT algorithm. The green grid that you are seeing on the photograph is the feature vector of the SIFT algorithm. It is a frame that has been created around a feature of interest, and the orientation of it follows the main gradient direction, and for each "box" in the grid, a set of descriptors is computed. The "size" of the grid reflects the scale of the descriptor created, and because those are feature vectors, they can be compared to other photographs where the same parameter may occur but at a different scale. Inside the grid, you will see arrows. Those arrows indicate the major directions of the gradient (on the grey-scale image)

2.3   Emergence and Timeliness of SfM-MVS in Geosciences

Only a decade ago, SfM-MVS was mostly unknown in the field of geosciences, but it has rapidly expanded into a mainstream technique, which is widely used in geosciences, including geomorphology. A guidebook and manual of "structure from motion in the Geosciences" [8] already exists providing an extensive background and just the right amount of technical details necessary for a geomorphologist who does not want to be swamped in mathematical considerations—I recommend geomorphologists interested in starting using the method and wanting more than what is in the present chapter to go and make it their next read. There are several methods and tools to produce point-clouds and even in the field of photogrammetry one could argue that other methods than SfM-MVS could be and are worth presenting. There is certainly no arguing around those issues, but as I had to make some choices, I am adding a few lines on the timeliness and the choice of the topic for this chapter, as I am convinced that it is a technique worth mastering for the geomorphologist.

First and foremost, geomorphologists—myself included —have been very quick to run towards new technologies like SfM-MVS, other photogrammetric techniques, different forms of laser scanning, etc., but arguably those are progress in scientific fields other than geomorphology. The products of this progress—like SfM-MVS—have been imported to geomorphology because they are mature enough to be used on complex "natural" objects and because they have passed the sole development stage, so that non-specialists could use them. Moreover, SfM-MVS is a method that is low cost and one just needs a camera and a reasonable computer/laptop. Also, it does not have a steep learning curve, so that it is really open to anyone interested in the application of the technique. Of course, I recognize that a lot of geomorphologists with technical abilities in either remote sensing, photogrammetry and other fields of engineering are capable to juggle a full set of more complicated methods, but only SfM-MVS (at present) allows geomorphologists to concentrate on questions and paradigms that define geomorphology without having to develop expensive budgets nor spend important amount of time in learning a new tool. SfM-MVS offers the right balance of finance/time investment for the geomorphologists to remain true to relevant paradigms.

As SfM-MVS is very low cost compared to any other high-resolution topography (HRT) and as it proposes a single workflow from the centimetre to the kilometre scale it has the potential to help the development of the method in most countries in the world. At present, HRT has seen a boom in international literature with a majority of publications from Europe, North America, North-East Asia (prominently Japan) and Oceania leaving behind countries that are slightly less wealthy. Although the scientific world is still dominated by post-colonial power relations, SfM-MVS HRT has the ability to see the increase of geomorphological studies in the economically less-favoured countries. With this concept in mind, in this chapter, we also provide some potential "alternatives" to the most expensive steps of the SfM-MVS.

2.3.1   Protocol Proposed for SfM-MVS Photogrammetry in Geomorphology (After James et al. [16])

As both laser technology and SfM-MVS have the possibility to generate 3D measures that can be used to understand geomorphic processes at the centimetre scale and even lower, the role of error in the measurement has become an important constraint. I would argue that a discipline that has emerged from field observation and conceptualization has not been fundamentally concerned with errors in measurements leading to uncertainty. However, point-cloud technologies for high-resolution topography have brought this issue to the fore. Indeed, geomorphologists who come from the field of surveying engineering or have a strong background in photogrammetry have been warning that in the face of the diversity of approaches that can be adopted when capturing SfM-MVS pointclouds, a protocol is necessary for those technologies become an integral part of the geomorphologist toolbox [16]. Furthermore, the variety of data acquisition processes also means that it is difficult to compare and assess the reliability of the calculations performed. In order to homogenize the field, James et al. [16], in a collective endorsed by the journal Earth Surface Processes and Landforms, have proposed a protocol to be considered as "best practice" for SfM-MVS used in geomorphology. Their article is extremely well constructed and clear, and I would like to refer the reader to this contribution, which can be broadly divided into the following essential steps:

Photographs and geometric data acquisition for SfM-MVS differs from traditional surveying techniques that aim to collect topographic data at a "similar scale", because it is a non-contact method and it is a passive remote sensing method, which does not involve reading the characteristics of an artificial signal return either. In other words, it is a lot of unknown to juggle.

Leaving aside the practical steps that need to be taken, data acquisition can be genetically defined as follows: (1) experiment design and preparation before you head into the field, (2) once in the field, preparation of metric elements that can be used to check the error of the model and also to calibrate the scale and the location of the survey; (3) take the photographs in a suitable fashion to afterwards being able to create a model (I suggest that you train and try first to have a feel about it), and make sure your acquisition method is adapted to the terrain/experiment you are working on and (4) do a quick check of your photographs on your acquisition system to see whether the photographs are what you need and whether there is any blind-spot you may have missed.

Data Acquisition Step by Step

Step 1—Experiment design and planning: Before you head out in the field or you start instrumenting your laboratory, the first question you must ask yourself is: what do you expect to capture and at what level of precision; i.e. are you investigating features that are relevant at a 10 m scale or do you look at features that need to be characterized at the metre or sub-metre scale. The second question that you need to ask yourself is: what is the error value that you could accept and that would not disrupt or disqualify the work you are attempting to achieve? If you are doing a one-off measure, or a series of measures, you then need to measure the error appropriately as well. Once you have answered those questions, you will know the distance you will need from the objects you are imaging (is 50 m distance alright, do you need to be only 50 cm away?). For instance, if you are imaging an area that appears to be about let's say 10 m 10 m in one single photograph and that you have a photo-resolution of 1000 pixels by 1000 pixels, you have one pixel that can image (at best) 1 cm, providing that the all surface is perfectly perpendicular to the orientation of the camera, but as it is never the case in nature, you definitely have an image with an accuracy that does not reach the centimetre. For instance, if you work on a mountain slope, or in an area with a coastal cliff and that you take a nadir photograph from a UAV or else, even if the horizontal plane (x, y) is 10 m × 10 m, the surface being slanted or even sub-vertical, you will have a reduction of the resolution that is a direct function of the angle of the surface to the x, y plane, which is horizontal.

After you have made decisions based on the characteristics of the field or laboratory data you want to acquire, it is time to start planning your trip in the field or your laboratory experiment. You will therefore need the camera with the pole to maintain it or the UAV and the camera to take the photographs, you may also want to pre-programme your flying path using one of the dedicated applications, make sure that you charge the batteries, bring as many spares as possible as well as the charger if possible. Data is often stored on a SD-drive or a mini-SD-drive. For a lot of UAVs, the photographs are stored on the aircraft, and each time you bring your UAV back to change the battery, you may want to change your SD-drive or download the data, especially if you fly in an area where you would not be able to go and recover the UAV if it was to go down. This is (from experience) a good way of not losing all the data. Finally, you will need the targets that you will spread in the field for tie points and for ground control. As for TLS, having another camera to document the process, as well as a hand-held GPS and a notebook, is essential. Compared to TLS, you will need to also have a way to record the exact positions of your targets on the ground. This can be using a GNSS in RTK or PPK mode, or it can be using a total station for instance. You will therefore need the equipment related. Please note that you may not need to do this each time, you may already have known points that you will use later on during the process, but if it is your first visit in the field, or if you don't have such data, referencing the targets on the ground will be an essential step.

Step 2—Data acquisition in the field: Once you are in the field with your equipment, you will need to first set up a network of targets on the ground that will act as tie points for your processing (accelerating processing and also joining different pointcloud together) and as GCPs to control the quality of your pointcloud. This is of course only if the surface you work on allows you to so (safety is always paramount). This network of points has to be installed first to appear in the photographs. If you can't access an area to place a target, make sure to use a total station or other measuring equipment to record the coordinates of features you will be able to mark on the photographs you are using to build the SfM-MVS model. I would strongly advise that you take photographs at the same time, as we all know that what seems obvious in the field is always so hard to remember or find in your photographs.

Step 2.1—GCP and tie points network set-up: The tie points are providing a scale and can be thought of as the boundary conditions of a differential equation, as it frames the point-cloud generated with SfM-MVS. In the field, it is therefore best to spread a network of tie points on the target object with variations in X, Y and Z. Furthermore, it is best to have tie points on the edges of the surveyed area as well as in the centre, to avoid the dome effect and to also avoid the edge and corner to unnecessarily rise or subside. On top of the targets that are used as tie points between your 3D SfM-MVS point-cloud and the field data, you want to also place at random a set of GCP targets that will be used to measure the error between the predicted data in your model and the data measured in the field. For beginners, please note that the targets need to be placed at the very beginning of the acquisition process, as they will need to appear in all the photographs you are taking.

Some tools to record the location of the targets: TLS, solid-state low-cost LiDAR sensor and total stations: Once you have placed the targets in the field, it is then time to record their position precisely using another tool such as a GNSS system for instance. Personally, I prefer to do this step, right after placing the targets, because for one reason or another (not enough sky, no line of sight), you may have to move your target from its original position. In such a case, you will need to take the UAV data again. It is therefore best to do it in this order: (1) place the targets, (2) record their location and then (3) do the UAV/camera survey.

Although tie points and ground control points targets are often the preferred system to tie SfM-MVS data with field data, it is, however, not always possible to go and place a target on the surveyed surface. For instance, if you are working on a cliff face in coastal areas or a rockwall in the mountains, placing your own security first might be the best option and instead of using targets, you may prefer to use another method. For this purpose, there are two possible approaches: (1) one that consists in generating a set of known points from recognizable features that you will be able to use in the modelling process as tie points and as ground control points (for instance, you can shoot with a total station a set of features on a sub-vertical wall and take photographs of where exactly you took the data. I often use the end of a crack for this purpose). The second method: (2) is to use another point-cloud that is being generated with no or less distortion using a TLS for instance, and then to map your new point-cloud on the first one. This method can be performed going back and forth between the SfM-MVS software and a software or a freeware allowing you to place two point-cloud over one another. You can then use the DEM of Differences (DoD) algorithm to compare how well your two point-clouds are matching.

Step 2.2 Taking the photographs: At this stage, your targets should be spread in the field, and you may already have collected their exact position using one method or another. It is now time to take the photograph (by hand maybe if you are in the laboratory or from a moving cart, or from a UAV, etc.). Taking the photographs for SfM-MVS requires you to follow a few rules. You cannot take the photograph randomly. They need to be taken in continuity and contiguous to one another, with an overlap of about 60% along the camera pathway, and with an overlap of >30% in the perpendicular direction to the camera pathway. The network of photographs needs to be forming a grid-pattern for computation purposes (Fig. 2.5).

If you take photographs of an object that appears to be a topography or close to a plane shape, you will be using a series of photographs aligned in a given direction and with a rotation at the end. From one photograph to another it is best to have slight angle variations (Fig. 2.6b) instead of fully NADIR photographs (Fig. 2.6a). It is important that the photographs overlap one another from the left and right side but also from the top and the bottom (Fig. 2.6c).

If you are taking photographs of a single object that is discrete, with faces in different directions, like a statue, a building, a column, you will want to rotate around the building by slowly increasing (or decreasing, depending on where you start) the altitude, a bit like the shell of a snail or the shape of a spring. This movement is occurring while your camera is of course looking inward.

The final step, once you have your photographs and your secondary data (GNSS or total station records of your targets) will be the creation of the point-cloud using one of the software available. In the next section, we are running through one solution with the Russian proprietary software Agisoft Metashape Pro (https://www.agisoft.com/), which recently the software Photoscan Pro. Another very similar solution, that is also convenient to use is the open-source Meshroom (France, Norway and Czech collaboration), based on the open-source algorithms of Alicevision (https://alicevision.org/#meshroom).

2.3.2 Example of a Step-By-Step Workflow with Agisoft Metashape Pro

The solution provided by Agisoft® (either Photoscan Pro or Metashape Pro) is certainly the most popular among geomorphologists, notably due to the low cost and the ease with which the software can be used. For this reason, I provide you with an example workflow (Fig. 2.7). This software is particularly attractive, because it proposes an intuitive interface (Fig. 2.7).

Running the SfM-MVS process in metashape is as follows:

(1) Import your photographs to the scene from the Workflow tab. You can import a full folder or photographs individually. Once this is done, you want to check the quality of the photographs, and right-clicking on the photograph, you can choose different options, such as looking for blur photographs, so that you can then decide or not whether you want to include them in the computation (blur photographs are the enemy of SfM-MVS).

(2) You can then key in manually sets of tie points (flag) or you can detect them, if you have used detectable markers (in the tools, you can print targets that the software will be able to recognize automatically when placed in the scene). The X, Y, Z values of the different targets can be keyed in manually, but you can alternatively load a csv file with the values. I suggest that you generate a csv file with one dummy value you keyed in to understand the format of the file. From this file, you will then be able to easily replace the dummy value by your own data.

(3) Once you have done this step, and that the targets are linked to the X, Y, Z values, you can go back to the workflow panel and run the SfM process, choosing the

Fig. 2.5 How to take photographs for a planar object (**a**), and for a discrete multifaceted object (**b**). Please note that the angle slightly changes along the path and that the photo-object vectors tend to converge

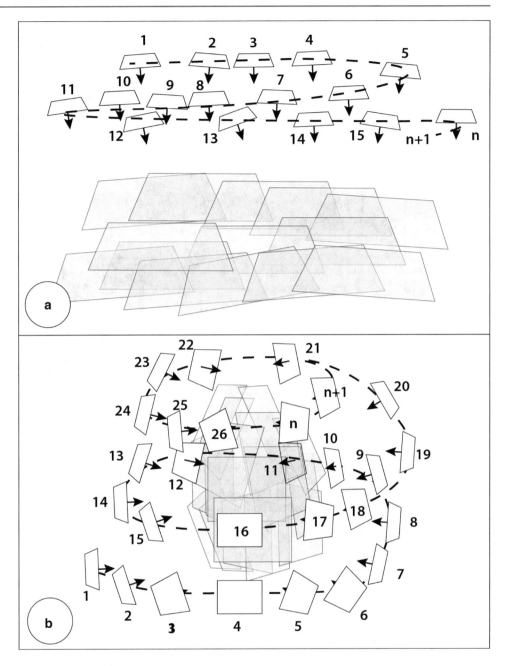

level of details you wish. Please note that in the options, choosing 0 as the point-limits will allow the pointcloud to grow without any limit.

(4) This first step will have generated the point-cloud form SfM, and you will be able to check whether the values in X, Y, Z you have entered from your field measurements, coincide with the data calculated. This will allow you to find targets that may have a problem or see whether there has been issues in the reconstruction process. If you open the "photographs" tab, you will also get information on the number of overlaps, etc.

(5) Using the SfM-MVS step, you can then open once more the "workflow" and run the "Densification", which is

the MVS step. This step can take a lot of time (24 h to weeks sometimes), and it will generate the dense point-cloud.

(6) Afterwards, if you are interested in generating a DSM and orthophotography and divide large dataset in smaller ones, it can all be done within the Metashape Pro software (one reason it is so popular I believe).

(7) Once you have the dataset in the format you want, you can then open the "file" tab and choose export and export the data in the format you want, as a text file or a LiDAR data file las or laz. Depending on what you want to do with your point-cloud at the processing stage, you will need to choose one format over another.

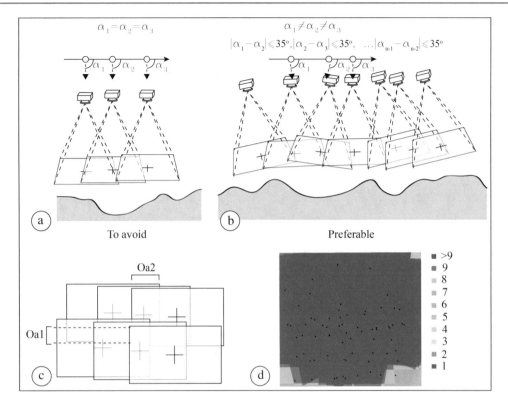

Fig. 2.6 Art of taking photographs of a surface for SfM-MVS. **a** It is not recommended to follow this pattern where all the cameras are aligned in the same direction and orientation, which makes the SfM process difficult. Instead it is recommended to modify the angle between each photograph, and so by not more than 35° (maximum) in between each shot, as shown in **b**. **c** The overlap in both directions between the photographs (arbitrarily named Oa1 and Oa2) should allow a single point-object to be seen from a minimum of three shots. Various overlap levels have been suggested by different authors, but from experience, it is best to capture as many photographs as you can with a maximum of overlap as you may have to discard some that are blur for instance and you don't want to have a gap in your dataset. **d** This is the overlap report provided by the software Metashape Pro® from Agisoft for the micro-topography experiment reported later in the chapter on volcanic structure. You can see that I have most of my areas of interest covered by more than nine photographs each time. This is certainly a bit of an "overkill", and the right balance is certainly a combination of the time available you have to take the photographs and the spatial extent of the field you would like to cover

2.4 From Quantitative to Qualitative: SfM-MVS "Photogrammology"

Before I close this first chapter, I wanted to say a few words on what I have named here photogrammology. The reason is to acknowledge that geomorphology can also be a qualitative science, and that SfM-MVS can provide a simplified qualitative or semi-quantitative first step, especially when instrumentation is not available to perform a full SfM-MVS analysis. I have named this process SfM-Photogrammology (meaning that it uses the language of photogrammetry "logos", but maybe does not provide a dataset that is fully reliable for quantitative measures) "metria" means measuring. The choice behind this last element is the recognition that SfM-MVS coming from the field of engineering is partly dominated by engineers who practise geomorphology. I believe that this move has helped geomorphology transition to a more quantitative science and open new possibilities in

geomorphological engineering, but it should not become the sole discourse of geomorphology, in places where qualitative developments also exist.

Qualitative or semi-quantitative geomorphology may not appeal to everybody, and it is often a question of culture and how the field has evolved over time. For instance, in Japan, where sciences only entered the country at the end of the nineteenth century from a need and an appetite for practical solutions and for engineering, physical geography and geomorphology have been strongly influenced by engineering, and the high-level mathematical entry examination at universities has pushed students and educators in this direction. Geomorphology is therefore integrated to civil engineering practices and "sabou" practices, which is sediment hazard management, traditionally in the field of agricultural sciences. In geography, Prof. Takashi Oguchi of Tokyo University has been instrumental in revitalizing geomorphology and physical geography by very early on, integrating Geographical Information Systems and technology

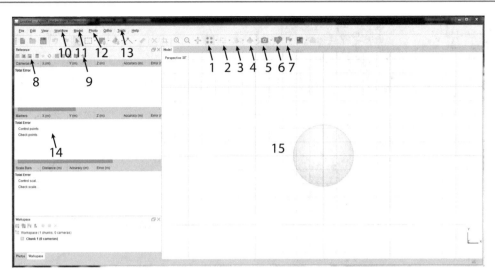

Fig. 2.7 Main window of Agisoft Metashape Pro (replacement of Agisoft Photoscan Pro). (1) Sparse point cloud; (2) Dense pointcloud; (3) Meshing; (4) Tiled model and photographs wrapped on the model; (5) shows the camera positions in the 3D view; (6) show shapes in the model; (7) show the control points and tie points you define; (8) import, save cameras and control and tie points coordinates; (9) Tools to modify the projection, the CRS, the theoretical accuracy of the coordinates; (10) the workflow where you will find all the different steps necessary to create your model; (11) where you can access the characteristics and options of your models; (12) the options of your photographs; and (13) the tools where you will find the automated targets and different options useful to construct the model; (14) the window where you will see the list of your photographs as well as the tie points and control points. (image of the GUI in 2020, one will note that this may change over time)

such as LiDAR point-cloud data. In this context, geomorphology is not a qualitative science, but in countries where it has been associated with geography at university, qualitative considerations such as a tradition of the field sketch for instance, are still persisting today.

Once again, SfM-MVS is definitely a tool developed for quantitative geomorphology and used properly [16], we can create datasets that are fitted for geomorphometry and multi-temporal evolution analysis. I think that SfM-MVS can also be used at the limit between quantitative and qualitative geomorphology, when you want to get an idea or a description of a morphology, or when there is nothing more in your backpack than your favourite camera, but not the finance to collect all the GCPs coordinates, etc. There are still steps for you to check on the quality of your 3D. It does not mean that doing semi-quantitative work needs to be totally written off. If you work at a local scale, from a hand-held camera, you can use a cheap 2$ (100 yen) measuring tape and you can set on to measure about every feature you can to constrain the size of the point-cloud you will create. If you work from aerial photographs, you can measure the size of buildings and features that you can recognize in the landscape or using a free tool such as Google Earth (although it is not perfect), and then you can calibrate the size of your model. For instance, I have used this strategy to easily stitch aerial photographs together, but I restricted the analysis to a 2D planform analysis with a low-resolution 2D surface analysis (Fig. 2.8). Obviously,

you won't have well-spread control targets and the control instruments are actually not professional-grade, but if you don't plan to use your data for quantitative purposes, and if you are clear about your use and the limitation of the dataset, you shouldn't stop yourself from using SfM-MVS. In such a case, one of your tasks will be to explain why the method is relevant and what are the limitations of the dataset that you created, and that you understand those limitations.

Besides the idea of recognizing that qualitative and semi-quantitative geomorphology should get a seat at the table, it also spans from the need to recognize that science should not be seen as a single direction and its paradigm vector should not be controlled by a set of countries that have access to the largest amount of funding, allowing to buy the necessary equipment. Sciences should also allow work to be done in other directions, providing that it is a progress. The French philosopher Michel Foucault argued that power and knowledge can't go without one another as power needs knowledge and knowledge generates power, but power gained from knowledge should maybe not be consolidated further by scientists when possible. For instance, calibrating SfM against GNSS-RTK, PPK or against TLS defies one of the reasons why UAV-based SfM-MVS has been used in numerous "developing nations": the limited amount of funding for research. Although one may argue that collaborations with well-funded research groups from other nations would solve this issue, it then becomes an ethical issue of neo-colonialism [13], denying

Fig. 2.8 Using SfM-MVS on historical aerial photographs scaled using buildings and elements that "remain" in the landscape today, a set of orthophotographs was created for the period 1941–2012. However, the quality and the number of images is often not enough to generate a proper 3D point-cloud and the data was used here in combination with old maps of the nineteenth century to look at the change of the Tamagawa River in Tokyo and in the present case the water surface in different reaches, under the influence of dam constructions and other management programmes

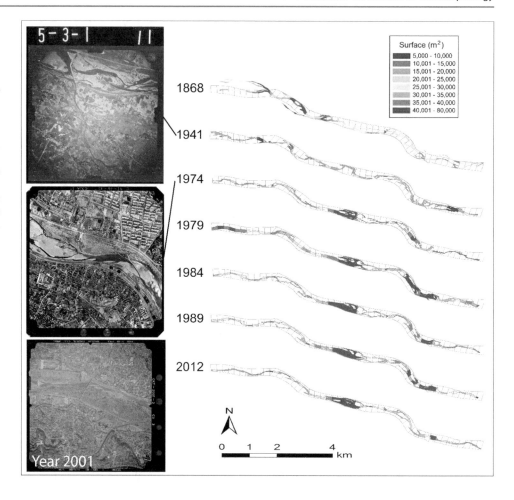

the right of intellectual autonomy and sovereignty of "less wealthy" countries.

In summary, "wealthy countries" researchers will certainly be happy to go and work with colleagues who can't afford the technology, but local researchers often won't share the same enthusiasm. And it is to fight those attitudes that a group of researchers from universities and research institutes from all around the globe have come together under the leadership of Assoc. Prof. J. C. Gaillard to write a manifesto: https://www.ipetitions.com/petition/power-prestige-forgotten-values-a-disaster. Although this manifesto, written and signed by researchers from The universities of Auckland (NZ), Lund (Sweden), Boulder Colorado (USA), the Philippines Diliman (Philippines), Longhborough (UK), NW (South Africa), Newcastle (Australia), Kobe (Japan), ICCROM (United Arab Emirates), UCL (UK), Copenhagen (Denmark), Shandong (China), etc., is intended for research in the field of disaster risk, it applies as well in the cross-cultural and international relations researchers have and the need to recognize the existence of power relations (notably artificially created by University rankings).

To a lot of readers, this subsection will come as a hair in the soup, but even for those who disagree with the usage of

SfM-MVS, hopefully it will provide them the chance to reflect on the fact that science is not a one direction vector.

Conclusion

This chapter has presented you with two of the techniques used to collect point-clouds for geomorphology: Laser technologies (ALS and TLS) as well as the SfM-MVS photogrammetric method. The laser technology is arguably the strongest one, as it allows "true measurement" with limited distortion compared to SfM-MVS, which can see emerging error if tie points and ground control points are not spread evenly across the surveyed space. However, SfM-MVS only comes at a fraction of the cost and allows work at the centimetre scale and lower, providing that you take the photographs close to the target. Laser technology, in such cases, will be limited by the instrument and the footprint of the laser.

In both cases, however, the secret of successfully collecting data lies in the preparation and planning, so that you develop the right methodology for your field and that you do not forget to take one or another dataset, that will be crucial to use the data. For SfM-MVS, the recently proposed protocol [16] is certainly a necessary read, to make sure that the process is streamlined, the data can be tested for reliability and to make sure the data can be compared with other dataset. As point-cloud data is increasingly accessible online, it is important that those point-clouds can be compared and reused by other researchers.

In the next chapter, I invite you to take the point-cloud you will have generated and see how you can process it to extract geomorphologic information.

References and Further Readings

1. Alizad K, Medeiros SC, Foster-Martinez MR, Hagen SC (2020) Model sensitivity to topographic uncertainty in meso- and microtidal marshes. IEEE J Select Top Appl Earth Observ Remote Sens 13:807–814
2. Balaguer-Puig M, Marques-Mateu A, Lerma JL, Ibanez-Ascensio (2017) Estimation of small-scale soil erosion in laboratory experiments with structure from motion photogrammetry. Geomorphology 295:285–296
3. Baltsavias EP, Gruen A (2003) Resolution convergence: a comparison of aerial photos, LIDAR and IKONOS for monitoring cities. In: Mesev V (ed) Remotely-sensed cities. Taylor & Francis, Hove, pp 47–84
4. Bistacchi A, Balsama F, Storti F, Mozafari M, Swennen R, Solum J, Tueckmantel C, Taberner C (2015) Geosphere 11:2031–2048
5. Brasington J, Rumsby BT, McVey RA (2000) Monitoring and modelling morphological change in a braided gravel-bed river using high resolution GPS-based survey. Earth Process Land 25:973–990
6. Brasington J, Smart RMA (2003) Close range digital photogrammetric analysis of experimental drainage basin evolution. Earth Surf Proc Land 28:231–247
7. Brasington J, Langham J, Rumsby B (2003) Methodological sensitivity of morphometric estimates of coarse fluvial sediment transport. Geomorphology 53:299–316
8. Carrivick JJ, Smith WM, Quincey JD (2016) Structure from motion in the geosciences. Wiley Blackwell, 197 p
9. Chandler JH (1999) Effective application of automated digital photogrammetry for geomorphological research. Earth Surfrace Process Land 24:51–63
10. Chandler JH, Ashmore P, Paola C, Gooch M, Varkaris F (2002) Monitoring river-channel change using terrestrial oblique digital imagery and automated digital photogrammetry. Ann Assoc Am Geogr 92:631–644
11. Chandler JH, Fryer JG, Jack A (2005) Metric capabilities of low cost digital cameras for close-range surface measurement. Photorammetric Record 20:12–26
12. Dietrich JT (2016) Riverscape mapping with helicopter-based structure-from-motion photogrammetry. Geomorphology 252:144–157
13. Gomez C, Hart DE (2013) Disaster gold rushes, sophisms and academic neo-colonialism: comments on 'earthquake disasters and resilience in the global north.' Geogr J 179:272–277
14. Gomez C, Allouis T, Lissak C, Hotta N, Shinohara Y, Hadmoko DS, Vilimek V, Wassmer P, Lavigne F, Setiawan A, Sartohadi J, Saputra A, Rahardianto T (2021) High-resolution point-cloud for landslides in the 21st century: from data acquisition to new processing concepts. World Landslide Forum 6:199–214
15. Heritage GL, Large ARG (2009) Laser scanning for the environmental sciences. Wiley-Blackwell Publisher, 278 p
16. James MR, Chandler JH, Eltner A, Fraser C, Miller PE, Mills JP, Noble T, Robson S, Lane SN (2019) Guidelines on the use of structure-from-motion photogrammetry in geomorphic research. Earth Surf Proc Land 44:2081–2084
17. Järnstedt J, Pekkarinen A, Tuominen S, Ginzler C, Holopainen M, Vitala R (2012) Forest variable estimation using a high-resolution digital surface model. ISPRS J Photogramm Remote Sens 74:78–84
18. Johnson MD, Fredin O, Ojala AEK, Peterson G (2015) Unraveling scandinavian geomorphology: the LiDAR revolution. GFF 137:245–251
19. Kociuba W, Kubisz W, Zagorski P (2014) Use of terrestrial laser scanning (TLS) for monitoring and modelling of geomorphic processes and phenomena at a small and medium spatial scale in polar environment (Scott River—Spitsbergen). Geomorphology 212:84–96
20. Kondolf GM, Angermeir PI, Cummins K, Dunne T, Healey M, Kimmerer W, Moyle PB, Murphy D, Patten D, Railsback S, Reed DJ, Spies R, Twiss R (2008) Projecting cummulative benefits of multiple river restoration projects: an example from the Sacramento-san Joaquin river system in California. Environ Manage 42:933–945
21. Kondolf M, Piegay H (2003) Tools in fluvial geomorphology, 1st edn. Wiley, Hoboken, 688p
22. Kondolf M, Piegay H (2016) Tools in fluvial geomorphology, 2nd edn. Wiley, Hoboken, 548p
23. Lane SN, Hicks DM, Westaway RM (1999) Monitoring riverbed topography by digital photogrammetry with particular reference to braided channels. NIWA technical report, New Zealand
24. Lane SN (2000) The measurement of river channel morphology using digital photogrammetry. Photogram Rec 16:937–961
25. Lane SN, Chandler J-H, Porfiri K (2001) Monitoring river channel and flume surfaces with digital photogrammetry. J Hydraul Eng 127:871–877
26. Lane SN, Westaway RM, Hicks DM (2003) Estimation of erosion and deposition volumes in a large, gravel-bed, braided river using synoptic remote sensing. Earth Surf Proc Land 28:249–271
27. Lin Y, Hyyppä J, Jaakkola A (2011) Mini-UAV airborne LIDAR for fine-scale mapping. IEEE Geosci Remote Sens Lett 8:426–430
28. Marks K, Bates P (2000) Integration of high-resolution topographic data with floodplain flow models. Hydrol Process 14:2109–2122
29. Marteau B, Vericat D, Gibbins C, Batalla RJ, Green DR (2017) Application of structure-from-motion photogrammetry to river restoration. Earth Surf Proc Land 42:503–515
30. Morgan JA, Brogan DJ, Nelson PA (2017) Application of structure-from-motion photogrammetry in laboratory flumes. Geomorphology 276:125–143
31. Pagan JI, Banon L, Lopez I, Banon C, Aragones L (2019) Monitoring the dune-beach system of Guardamar del Segura (Spain) using UAV, SfM and GIST techniques. Sci Total Environ 687:1034–1045
32. Pavlis T, Mason K (2017) The new world of 3D geologic mapping. GSA Today 27:4–10
33. Pearson E, Smith MW, Klaar MJ, Brown LE (2017) Can high resolution 3D topographic surveys provide reliable grain size estimates in gravel bed rivers? Geomorphology 293:143–155
34. Persendt FC, Gomez C (2016) Assessment of drainage network extractions in a low-relief area of the Cuvelai Basin (Namibia) from multiple sources: LiDAR, topographic maps, and digital aerial orthophotographs. Geomorphology 260:32–50
35. Petrie G, Toth CK (2008) Introduction to laser ranging, profiling, and scanning. In: Shan J, Toth CK (eds) Topographic laser ranging and scanning, principles and processing. CRC Press, Boca Raton, pp 1–27

36. Petzold B, Reiss P, Stossel W (1999) Laser scanning—surveying and mapping agencies are using a new technique for the derivation of digital terrain models. ISPRS J Photogramm Remote Sens 54:95–104

37. Rey-Otero I, Delbracio M (2014) Anatomy of the SIFT method. Image Process Line. https://doi.org/10.5201/ipol.2014.82

38. Sanhueza D, Picco L, Ruiz-Villanueva V, Iroume A, Ulloa H, Barrientos G (2019) Quantification of fluvial wood using UAVs and structure from motion. Geomorphology 345:106837

39. Saputra A, Rahardianto T, Gomez C (2017) The application of structure from motion (SfM) to identify the geological structure and outcrop studies. In: AIP conference proceedings, vol 1857, p 030001. https://doi.org/10.1063/1.4987060

40. Scherz JP (1974) Errors in photogrammetry. Photogramm Eng

41. Schrott L, Otto J-C, Geilhausen M (2013) Fundamental classic and modern field techniques in geomorphology: an overview. Treatise Geomorphol 14:6–21

42. Schwendel AC, Milan DJ (2020) Terrestrial structure-from-motion: spatial error analysis of roughness and morphology. Geomorphology 350:106883

43. Shan J, Toth CK (2008) Topographic laser ranging and scanning—principles and processing. CRC Press, Boca Raton

44. Speitzer G, Tunnicliffe J, Friedrich H (2019) Using structure from motion photogrammetry to assess large wood (LW) accumulations in the field. Geomorphology 346:106851

45. Vazquez-Tarrio D, Borgniet L, Liebault F, Recking A (2017) Using UAS optical imagery and SfM photogrammetry to characterize the surface grain size of gravel bars in a braided river (Veneon River, French Alps). Geomorphology 285:94–105

46. Wang C, Yu X, Liang F (2017) A review of bridge scour: mechanism, estimation, monitoring and countermeasures. Nat Hazards 87:1881–1906

47. Westoby MJ, Brasington J, Glasser NF, Hambrey MJ, Reynolds JM (2012) 'Structure-from-motion' photogrammetry: a low-cost effective tool for geoscience applications. Geomorphology 179:300–314

48. Westoby MJ, Lim M, Hogg M, Pound MJ, Dunlop L, Woodward J (2018) Cost-effective erosion monitoring of coastal cliffs. Coast Eng 138:152–164

49. Yano A, Shinohara Y, Tsunetaka H, Mizuno H, Kubota T (2019) Distribution of landslides caused by heavy rainfall events and an earthquake in northern Aso Volcano, Japan from 1955 to 2016. Geomorphology 327:533–541

Abstract

This chapter starts with a presentation of the common pointcloud formats used in geomorphology and civil engineering, as well as a rapid presentation of three freeware and software (with free academic licence) available to read and process pointcloud data. The chapter then explains the registration process of pointclouds and the classification of points including some of the most commonly used algorithms (slope-based and modified slope-based filters, the progressive morphological filter, the low-topographic gradient filter). The chapter then presents some new ideas for processing pointclouds in geomorphology and how data can be locally vectorized to decrease the processing time. After this first section on "one pointcloud", the chapter then rapidly presents a comparison of two pointclouds running in the CloudCompare software (this is a topic that has been well-developed by other geomorphologists and thus the section is not developed further). The chapter then continues on the conversion of data format. As numerous geomorphologists will work in a GIS environment that deals with shapefiles and gridded data, the chapter then discusses the limitations and advantages of converting data formats (TIN and grids) before presenting the common measurements and derivatives that can be taken from converted pointclouds. This section also has worked examples using the R language and the LidR library to extract gridded data from point-clouds, etc. The chapter is then finished with quality measurement and error analysis of the pointclouds, including RMSE, MBE, MAE, SDE and SE, with once again worked examples in R.

Learning Outcomes: Hands-On Processing and Know-How

– Know the different processing steps of a pointcloud generated in the field;
– Understand the main different filtering algorithm;
– Know the main pointcloud data formats used in geomorphology;
– Be able to calculate the error from a pointcloud dataset;
– Understand the full process from data collection (Chap. 2) to processing to get a dataset ready for geomorphological analysis;
– Know the main derivatives generated from pointclouds, DEMs and DSMs.

Objectives: From Raw Pointclouds to Derived Products

The objective of this chapter is to provide you with the necessary information to bring your raw pointcloud, just "out of the field" into a dataset that you will be able to use for geomorphological analysis. The last part of the chapter also presents some of the most common derivatives.

First, I invite you to look at the different types of point-cloud formats, before addressing the question of registration, which will help you align and eventually scale your point-clouds and place them in a geographic reference system (very often your pointcloud will be in a local coordinate system). Once you have this registered pointcloud, land cover, the ground and other errors are still part of your dataset, and you will need to filter and classify your point-cloud based on the morphology of the point-cloud or based on the characteristics of each point, as you would do with traditional remote sensing. Then, afterwards, I am presenting you two filtering and deconstruction algorithms that are still work in progress, but which modify the filtering angle, as I argue that we need to stop seeing pointclouds as "point—clouds", because they are the end of a vector between a recording (and sometimes emitting) apparatus, and the vector between the two is holding data on the location of empty space. Finally, I ended the chapter with data conversion to other 2.5D and 2D surfaces, as well as their derivatives.

C. Gomez, *Point Cloud Technologies for Geomorphologists*,
Springer Textbooks in Earth Sciences, Geography and Environment,
https://doi.org/10.1007/978-3-031-10975-1_3

Who is the chapter for? This chapter is also intended for undergraduate and graduate students who are learning point-cloud technologies such as structure from motion and LiDAR, with a background in geosciences, either geomorphology, geology, geography, geophysics, etc. Scientists and more experienced researchers who also need to process data as part of their work and need a few steps of know-how.

Who is the chapter not for? This is not a comprehensive course or ersatz to a course on pointcloud processing, and it only presents the main steps and algorithm. It won't provide you with case-specific issue resolution (which may be beyond the scope of any book anyway).

Introduction

Butler et al. [4] start their abstract with "[…] large point cloud datasets become ubiquitous in the Earth science community, […]", and it thus appear essential for the geoscientist to know how to handle and process such dataset, even if he or she does not become a full-time "point-cloud processing expert". This sentence crystalizes particularly well the present state of geoscience with regard to point-cloud technologies, but also the need to develop further processing algorithms to extract the most of these datasets.

Even if laser scanners are still coming with a hefty price-tag, the huge amount of data collected by governmental agencies (in yet a relatively limited number of countries) and made available to the scientific community has shifted the geoscientific methodological paradigm from being able to "get data" to "how to deal with so much data and making sense of it". And even in areas where laser data is not available, low-cost SfM-MVS photogrammetry can often fill the gap.

At this stage, when you are reading this book, you should already be familiar with the data acquisition techniques using laser scanning and SfM-MVS photogrammetry, and you may already have a dataset on hand that you want to process, and you may not be quite sure what is the next step to take, and solving this issue will be the goal of this chapter.

3.1 Point-Cloud Format, Visualization, Classification and Division

If you have been, or if you are going through a traditional geomorphology cursus, it is most likely that you have learned some remote sensing and some GIS, so that one of the question that you may have is, how do I get from the data I have collected in the field to a dataset, which I will be able to use in one of my usual software environment, and what are the different formats the dataset might be provided to me in (if for instance you are receiving a point-cloud from a LiDAR survey for instance).

But before one thinks about converting the original dataset, it is first important to identify and separate objects of interests, especially because pointclouds hold information that may disappear at a later stage during the data format transformation. This ability to extract object and information from pointclouds using classification techniques is, most contemporary geomorphologists will agree, what has realized the full potential of laser scanning and other High-Resolution Topography (HRT) acquisition methods [3]. Furthermore, these two authors have recognized that classification is not a benign process, because of "(1) the 3D nature of the data as opposed to the traditional 2D structures of digital elevation model (DEM); (2) the variable degree of resolution and completeness of the data due to inevitable shadowing effects; (3) the natural heterogeneity and complexity of natural surfaces; and (4) the large amount of data […]" (from the introduction of Brodu and Lague [3]). It is therefore essential to be able to read, use and translate the different types of datasets from one format to another without losing in quality, quantity, nor modifying the information. Knowing the different files' types and their limitations, you should then be able to use the most appropriate file type depending on your goal.

3.1.1 Point-Cloud Formats and Extensions

If you created a point-cloud from SfM-MVS or from laser scanning (previous chapter), it can be saved in a variety of formats. Although laser scanners will often generate binary or protected proprietary formats, SfM-MVS and most of the software used with the scanners will allow you to export your data in one or several of the following formats: Table 3.1. For instance, if you are using SfM-MVS and Agisoft Metashape Pro, you will be offered the following options: Adbove 3D pdf (.pdf), ASCII pts (.pts), ASPRS LAS (.las) and compacted format (.laz), ASTM e57 (.e57), Autodesk DXF (.dxf), Potree (.zip), Standford Ply (.ply), Topcon CL3 (.cl3), U3D (.u3d), Wavefront object (.obj), XYZ points (.txt). Another SfM-MVS application may create different file types, but as a geomorphologist those are the major formats you will use as a geomorphologist (Table 3.1).

Once you have opened one or two of the point-clouds that you have received or created, you will realize that during field acquisition, the pointclouds can be separated into different files, and that you may need to combine them, and eventually change from the local coordinate system of the laser scanner or the local scale used for your

Table 3.1 Major point-cloud data formats and their sizes as well as seamless (drag and drop) opening capacities for CloudCompare, Autodesk Recap (free for students for non-commercial use) and with the LidR library in R (added here because numerous examples are using this library in this book)

File type	File size (kb)	Seamless import in CloudCompare	Seamless import in autodesk recap	Seamless import in R with the LidR library
Adobe PDF	2203			
ASCII Pts	9276	o	o	
ASTM e57	1810	o	o	
Autodesk DXF	16,256			
LAS pts	4404	o	o	o
LAZ	2097	o	o	
Potree	1726			
Stanford PLY	2541	o		
Topcon PCL3	5250		o	
U3D	2200			
Wavefront Obj	11,515	o		
XYZ textfile	8607	o		

The file size in the table refers to the file structure without any data you may have recorded (formats available in 2020)

photogrammetric data generation, to a global coordinate system. And to do so, you will start with pointcloud registration.

3.1.2 Pointcloud Registration

When you generate a point-cloud from either a laser scanner or a topographic station of other sort, or from cameras with a set of ground control points (GCPs), it is most likely that you may not be able to take all the different elements of the scene from only one vantage point. In such a case, you will need to stitch your different points with one another. Often, you will be able to have a RTK GNSS or PPK GNSS on top of your laser scanner, so that your different point-cloud may almost match one another, and you may only need to optimize the alignment of the point cloud. In other cases, let's say if you take a TLS or SfM dataset under dense vegetation cover without good GNSS data. In this case, you will need to rely on a set of known points (for instance, man-made targets), which you will be able to see in the different scenes. The process of aligning the different point-cloud with one another is known as point cloud registration. This process is essential to create a full scene, but also to compare two scenes that may have changed over time. This is also the essential step in the method I have proposed to measure sediment, rock and soil density from SfM-MVS.

Pointcloud registration is divided into two steps, first a coarse registration that "roughly" aligns two point-clouds, followed by a fine registration. If you have downloaded one of the freeware used across this book, CloudCompare, you will find in the top horizontal menu, two buttons next to one another: "Align two pointclouds by choosing four pairs" and "Finely register already to roughly aligned entities". To use these options, you just need to use one already registered point-cloud and a second one that you will adjust to the first. If you work with a laser-based point-cloud you can assume that the distances between the points, the geometry is true, but when you work from point-clouds issue from SfM-MVS photogrammetry, it is possible that deformations may have gone through the cracks, even if you were careful in your process. In such a case, when you register your pointclouds, you want to make sure to also allow CloudCompare to scale the point-clouds, on top of being able to translate and rotate it.

A common method used for point-cloud registration is based on the reduction of the mean square error between two point-cloud, and the algorithm based on this method is usually named the Iterative closest point algorithm. Other methods use the same concept, but by comparing surfaces derived from the point-clouds for instance. CloudCompare uses the Iterative closest point. Although it is an effective method for medium size point-clouds, it has limitations when dealing with large point-clouds, which can slow the machine very significantly.

3.1.3 Classification of Points/Selecting Points for Computing

At this stage, you could have your dataset in the format of a point-cloud (collected using one of the different techniques explained in Chap. 2), aligned and eventually you may have done some filtering before even the registration of your pointcloud. As the point-cloud regroups ground, vegetation, building returns of your laser scanners, or of your point-cloud generated by photogrammetry, you will need to divide this point-cloud into different types of entities. Very often, the geomorphologists will want to separate the vegetation from the ground for instance, to retrieve the geomorphological information, but also sometimes to extract information about the ground nature based on the vegetation density, height, etc. For this step, you will need to use classification and filtering procedures.

This step can be realized using two different types of software and programming environments. It can be done with a true 3D point-cloud computing environment (such as CloudCompare), or using the point-cloud in a GIS environment using a combination of DSM and DEMs (in QGIS for instance).

Although it might be tempting to use the data generated "as is", this step is key [6] in the creation of a reliable dataset (you don't want millions of points on the vegetation, when you are interested in the ground for instance). And the task would be easy if both the vegetation or buildings were spaced regularly and with similar characteristics, so that the all topography could be brought back to a horizontal plane, from which we could extract elements based from their altitude, but as we all know topographic change, spatial variability in the land cover are all here to make our tasks more difficult, especially when there is no guarantee that within a spatial window, there are any return from the ground, it might just all be thick vegetation or a building roof. Consequently, scientists and engineers have developed different algorithms to filter point-clouds. A typical example of such a process will result in the stepped differentiation of the point-cloud (Fig. 3.1).

Morphological filters are algorithms based on the selection of a geometric value in a moving window (like a moving average), in order to extract certain features from the point-cloud, like removing vegetation from the ground data, or a building from the street level, etc. Early filters and still the majority of them consider the point to be filtered and its geometric relation to other points [15, 21].

– Slope-based filter [21]

The slope filter is an algorithm that looks for the lowest point within a region defined by the user (or the algorithm), using the slope or curvature of a cone, whose lower tip (the lower point) is located on the ground. The points located on or in the vicinity of the conic search surface will accept different points depending on the curvature of the conic surface. Comparing different window sizes with the slope-based filter, Asal [1] has identified that for a point-cloud generated

over a surface ranging from 165 to 292 m (based on the airborne LiDAR DSM), a window size of 3×3 m (horizontal size) would generate a mean altitude of 263.8 m, while the original point-cloud was at 265.2 m, this mean value changes up to 262.4 m with a window size of 41 m 41 m. The filtering algorithm is therefore significantly changing the result extracted from the LiDAR dataset. However, it does not mean that one method is more accurate than the other. As the windows grow wider, we lose horizontal spatial accuracy, but without any prior assumption, it is difficult to tell whether one method is better than another. As a rule of thumb, sharply topographic contrasted regions such as mountains and volcanoes will need a smaller window to account for the local topographic "jumps" while it may certainly be more efficient to work with a broader window in plains with small topographic gradients.

For the cone method, it can be expressed as follows [19]:

$$\forall P_j \in A : hp_i - \Delta h\big(d\big(p_i, p_j\big), m\big) \leq hp_j \qquad (3.1)$$

where P_j is a point in the dataset within the window excluding the point being examined; hp_i and hp_j are the heights of the i points in the window and of the j point being examined, m is the absolute value of the cone's gradient, and the negative value of m is the height of the cutoff plane, while A is the set of points to be filtered out.

– Modified slope-based filter [19]

One of the variants of the slope-based filter is the adaptive filter proposed by Sithole [19], because he noticed that if the topographic slope gradient is important, then the slope algorithm loses its shine. In his new equation, he replaced the general "m" by a value varying with the sampling window, in such a way that rapid topographic change could be accounted for. This type of filter can be grouped in the "adaptive", types of filter, and so is the one by Zhang et al. [23].

– Progressive morphological filter Zhang et al. [23]

After loading the raw airborne LiDAR data and after having generated the minimum surface grid, Zhang et al. [23] applied a morphological filtering to divide filtered terrain surface model from anything else that can be classified as "nonground point". Then, the size of the filter was increased to determine the elevation difference thresholds between "nonground point" and the terrain surface. Then, if the filter at a given size is larger than the maximum filtering window, they generated the digital terrain model. If the size of the filter could still be increased, they iterated once more over the dataset. Zhang et al. [23] proposed that for the algorithm

Fig. 3.1 Classification example of dataset generated from an ALS (airborne laser scanner) data

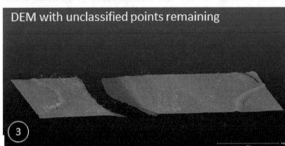

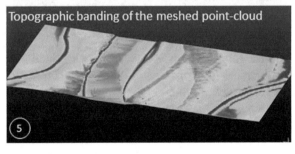

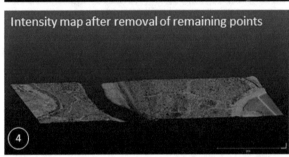

LiDAR data comes with the earth surface, most often of interest to geomorphologist and the vegetation on top that most of us just can't wait to get rid of. Happily liDAR data provides an easy solution to do so. In the present case, using the Cloud Compare freeware, we opened a .las file of the Missouri River in USA (1) and then extracted the vegetation (2). There are two easy way to remove vegetation, the first one is to work from the return number, the element the further away from the sensor being often the ground. Another way is to work from the return intensity, which is a function of the nature of the surface (vegetation against the ground, wet ground, etc.). This is the option we chose to take off the vegetation (2) and keep the ground (3). There is certainly no silver bullet method in remote sensing, and it is the same here. If you squint on (3) you will find a remaining hallow of points above the surface, that are remains from the vegetation that were mis-classified. To eliminate those, we use a nearest neighbor and density function removing the "outliers" that are not part of a dense population. Once this is done we then have the ground level (4) Of course it is a theoretical ground level and looking at the intensity map, we can discover that the ground is made of very different returns once again. As a geomorphologist, this is the data I am interested in, I can remain at this level of refinement and extract topographic data, profiles, the DEM... (Fig. 5) or I could further look at the ground with different density of vegetation cover, the presence of water, etc.

For this section of the Missouri floodplain, extracted from one of the LiDAR dataset, it took less than 30 minutes to go through the process (and record the different screenshot for this figure) on a laptop. It is therefore a very efficient way to obtain a DEM from LiDAR data.

the window size increased (in vertical pixel size) could be increased linearly between each time step, or alternatively the window size could be increased exponentially. In their publication, they provided the algorithm to implement the method, and if you use the OpenCV library, there is an implementation in C++ of the algorithm as well.

– Low-topographic gradient filtering: working with coastal mudflat [17]

The algorithm proposed by Pinton et al. [17] was specifically adapted to coastal mudflats, where the local topography shows very little variation. The authors divided the LiDAR dataset into 0.4×0.4 cells, and the minimum elevation was used as the best estimate of the ground under vegetation. They compared field data and laser data on 1.2×1.2 m grids, above which variation in the topography was introducing error.

The main steps of the algorithm are:

(1) Define the points $PC_{n,e}$ as the pointcloud subset in cell n,e of dimensions 0.4×0.4 m
(2) Compute the minimum elevation Z_{min} of $PC_{n,e}$
(3) In a 3×3 m plot centred on the cell defined in (1), they computed a regression surface $F(x, y)$ (for those not familiar with the procedure, it is like a simple regression line but with one extra dimension). They modelled the surface using a second-order polynomial regression, which the authors formulated as:

$$F(x,y) = \beta o_{n,e} + \beta 1_{n,e}x + \beta 2_{n,e}y \\ + \beta 3_{n,e}x^2 + \beta 4_{n,e}y^2 + \beta 5_{n,e}xy \tag{3.2}$$

where $\beta o_{n,e}$ is the intercept of the surface with the vertical axis passing through the centre of (n, e)

(4) The authors then computed the vertical distance between the points and the vertical projection on the regression surface, and using this step, the authors filtered the general surface variation, so that they could also work on the characteristics (micro-variation) that are attributed to the vegetation. This is a similar process to running a moving average window and then subtracting the moving average window from the original dataset.

- CANUPO: Filtering for complex geomorphologic surfaces [3]

The two French researchers N. Brodu and D. Lague are well known for the filtering algorithm they have generated to classify points from complex surfaces, separating vegetation from the ground (in horizontal and sub-vertical positions). Indeed, as HRT acquisition methods such as TLS are providing "non-aerial" angles to the surface topography, and so with high densities. Their algorithm has become very popular, because it is fully integrated in the CloudCompare open-source software and it does not require specialist training to be used.

Their algorithm investigates the surrounding points of a given point using a "neighbourhood ball", which is a sphere that can be scaled and that compares the neighbouring points at different scales.

– Computing the filters and other differentiation methods

At this stage, you certainly understand that before proceeding to the assignment of a 3D point to a class or a group of any sort, the question of the organization and selection arises, with concerns of accuracy, repeatability and also computing time. Furthermore, the structure of your pointcloud will also influence the process. For instance, when point-clouds are structured, they can be treated as gridded data, but when point-clouds are unstructured (like it would be for SfM-MVS and laser scanning data), the data processing presents a different set of challenges, with variable densities in space, etc. It is therefore necessary to reflect on these variations before you even start to apply one filter or another.

There are two main ways to run classifiers, unsupervised classifiers that will work through a point-cloud following a set of rules (expressed mathematically or statistically), that's why they are also called ruled-based classifiers or supervised classifications. The second is usually more "powerful" as it allows the user to modify and combine classification rules. However, it is more time-consuming than unsupervised classification as it necessitates the construction of a rule from a small (manageable) point sample, before it can be tested, and upon success, it can be extended to the all or several point-clouds.

Indirect methods or methods that are driven by the data are classification methods using machine learning. Those methods are rather manpower-intensive in the first states, because they need the training of the model by the creation and the modification of the classifier. This kind of classifier is not based on the application of a mathematical rule to the dataset, but the investigation of similarities between the presented point-cloud and the trained point-cloud.

3.1.4 Forward Outlooks for Filtering and Classification

Numerous point-cloud filtering and classification techniques used in geomorphology have been adapted from satellite remote sensing and from civil engineering applications, where laser scanners for instance appeared first. As the mapped structures are often simpler in geometry than the natural environment, the filtering and classification methods have seldom been developed with geomorphology in mind, and there is certainly plenty of opportunities for the geomorphologists interested in point-cloud technologies to develop original filters and classification methods that are best adapted to the discipline, such as the work done on coastal mudflats by Pinton et al. [17].

Pointcloud Processing as Vectors

Both for Sfm-MVS- and LiDAR-generated point-clouds, filters and classification methods use the target objects, but the relation of these reflector points and the camera or the laser scanner position and geometry could certainly be explored and exploited further.

One novel direction is to stop viewing point-clouds for what they are not. Indeed, point-clouds are deceivingly "not point-clouds", they are one end of a vector linking one or several cameras' pinholes or a laser emitter and receiver, either ALS or TLS. This results in a spatial structure that is a set of vectors defining an empty space where they flow through first and last returns, space in between which a medium that is either porous or the space in between two or more reflective membranes exists, like under a canopy structure for instance. There is therefore this empty space, which can be defined as empty because it is the space the laser had to cross before encountering one or several reflective surfaces (Fig. 3.2). If we can remind ourselves of the cone or adaptive-cone filter, we understand how much simpler the filtering process can become if we use vectors through free space landing on a surface, rather than just the point on the surface to be defined. It allows us to define space not only from the characteristics of the object itself, but also from the volumes where the objects are not located.

Adaptive Cylindrical Coordinates' Algorithms for Vectorization

Another direction that can be explored is the transformation of the dataset from a 3D dataset to a 2D series that can be worked in the frequency space. This sequence is similar to the slope method, where for each point of the point-cloud, surrounding points geometry is used. In the present case, we can define the position of a point on a line as being the distance to a selected point. Once you have extracted all the points in the vicinity of a given point, and you place them on a horizontal scale, which is defined by the distance to the selected points, you can then add on the vertical axis, any attribute of the point (intensity, RGB value, altitude), and this creates a 2D synthetic signal of the 3D point-cloud you are analysing. For each point, you will then create a 2D synthetic signal. If you then work on this signal using Fourier transforms or wavelet analysis, you will then have sets of signals that you can use to differentiate different "zones" in your pointcloud for instance. A wavelet decomposition will allow you to filter different values at given scales, and each time it will be in relation to one selected point in the point-cloud, in such a way that you are creating sets of attributes for each point in the point-cloud, which relate them to the surrounding points, based on any type of frequency or geometric function.

Fig. 3.2 Spatial structure of a point-cloud linked to the emitting and recording unit (in this example a TLS unit). **a** TLS and the representation of the laser scan with the laser path and the first (circle) and second (star) return. This can then be translated into a division of space (and not only a pointcloud) as in **b** with the supposed free space, obstructed space (this can be layered in density) and the ground space. Such an approach can be a good help in spatially structuring the pointcloud

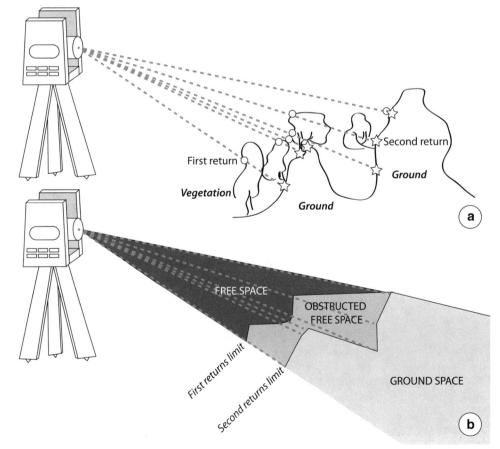

For instance, you can think about the present simple algorithm through this example. First, you need to vectorise the 3D point-cloud based on distances. This allows to assign various values on the structure in the neighbourhood of a point to this point, creating a richer point-cloud.

First, we want to transform the data centre $x_i = 0$, $y_i = 0$, $z_i = 0$, where the subscript "i" symbolizes the subscript of a i-point in a n series, so that the new point cloud P^n is:

$$P^n_{[x_n,y_n,z_n]} = \sum_{i=1}^{n} \sum_{j=1}^{n} P_{[x_i,y_i,z_i]} - P_{[x_j,y_j,z_j]} \qquad (3.3)$$

Because we are interested in the distance from a point and the positions of points to one or several others without gridding the data (which results in a loss of data accuracy), it is easier to work from polar and cylindrical coordinates. In the present case, we are using polar coordinates, modifying the new point-cloud P^n using the simple coordinate transformation formula:

$$P^n_{[p,\theta,z]}\left[\rho_i = \sum_{i=1}^{n} \sqrt{x_i^2 - y_i^2}\, \theta_i = \sum_{i=1}^{n} \tan^{-1}\frac{y_i}{x_i} z = z\right] \quad (3.4)$$

so that a point-cloud can be expressed to a point by a horizontal distance, a vertical distance and orientations separately (Fig. 3.3). From this construction of the data, you can then assign, calculate, recalculate any Z value of x.

You can also extend the work you are doing on the point-cloud by simply using the vectorised version, as it is more "computation-time" friendly. Indeed, in a 3D environment (in Cartesian coordinates for instance) the measure of distances in a buffer is more time-consuming than from a

vectorised data (like the one in Fig. 3.3). It is then possible to sum the number of points within a given horizontal distance or real distance (combining p and z), as extremely simply the sum of all points at a distance:

$$\sum_{i=1}^{n} \rho z_i \quad \text{for a given } \rho_n \qquad (3.5)$$

And the sum you are applying can be any attribute you may be interested in, you can also in the same way (just to give another example) calculate the variance, so that for a given point, and a defined space for a given ρ_n then the variance S^2 around this point can be expressed in cylindrical coordinates as:

$$S^2 = \frac{1}{n-1} \sum_{i=1}^{n} \left(z_i - \underline{z}\right)^2 \qquad (3.6)$$

from which the square root can be calculated to give the standard deviation. Working from vectorised point-cloud is not only "computing time friendly" it also allows you to fold back on a lot of the elementary statistics we are all very familiar with, and that R or Python with the Numpy and the Pandas library can handle very effectively.

In your steps towards producing a dataset for any of your geomorphological application, once you have filtered the data, you can certainly then use one of your favourite application to convert the point-cloud into raster or vector 2.5D that can be read by a GIS environment and from which mapping and other activities can occur. But, before folding back to your favourite GIS environment, you want to consider what it is that you are planning to do with your data.

Fig. 3.3 Representation of the points along the horizontal distance to a given point with Z vertical. Working with a set horizontal allows for the creation of densities, and colours or return values around each point, and for each origin point, the concentration of points in different buffers, so that you can define a point in its environment. This allows for a vectorization of otherwise data in a 3D space

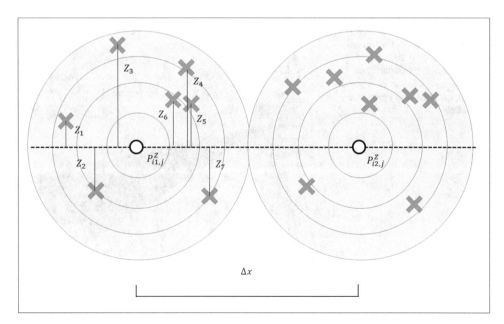

3.1.5 Comparison of Two Pointclouds

If you are thinking of doing time-series analysis with different point-clouds recorded at different times, it is certainly better to refrain from doing it in GIS. Indeed, GIS is the traditional environment that geomorphologists have used to compare different surface data over time and attempt to define erosion or deposition or landform changes, but as GIS is dealing with data in 2.5 D (i.e. projected on a plane or the Earth surface plane), you are then losing some of the information that your full 3D dataset has. This is particularly true when you are comparing two topographic datasets. In a GIS environment, the comparison between two surfaces will always occur vertically, because the sub-vertical faces are not represented in the system. If a cliff retreats by one metre and the cliff itself is 20 m high, the GIS system will show you a 1 m large strip with 20 m topographic change. In a full 3D environment, you can also measure the distance between the two topographies in the direction of the wall, so that you will be able to see the face of the wall retreating by 1 m.

From a more practical perspective, a full 2.5D and a full 3D are the two common end-members to compare point-clouds (except vectorization), but researchers have been mixing approaches and elements in 2.5D and 3D. For instance, comparing different point-clouds on a limited experimental surface, Rowley et al. [18] used a derived TIN surface at the base data from a LiDAR dataset at 1 point per 0.25 cm linear density spacing. From this surface, the authors compared the original LiDAR and the SfM point-clouds. They measured a RMSE at 3.5 cm resolution was, respectively, about 0.3, 0.63 and 0.64 cm. Their work over cobble, gravel, hummock, block and dunes has shown differences in the residuals, which vary with the type of surface when comparing SfM and LiDAR, suggesting that the nature of the target has as much importance as the technical set-up.

This is another element of importance if you are comparing pointclouds in geomorphology. Make sure that you reflect on what is the surface you are comparing, what is their roughness, you can also perform tests in the laboratory to try your method (Fig. 3.4). The sample from a set of experiments I conducted at Canterbury University in New Zealand at the Geography Department uses a gridded floor, on which sediments mixed with different water contents and overflowing water were tested. As I measured the mass of every single element involved, I was testing the volume—mass relation and also how different volumes of sediments with different moisture contents interact with one another. In the same manner, you can use point-cloud acquisition and analysis to derive properties of your experimental materials and how much you can also capture with the method you are using.

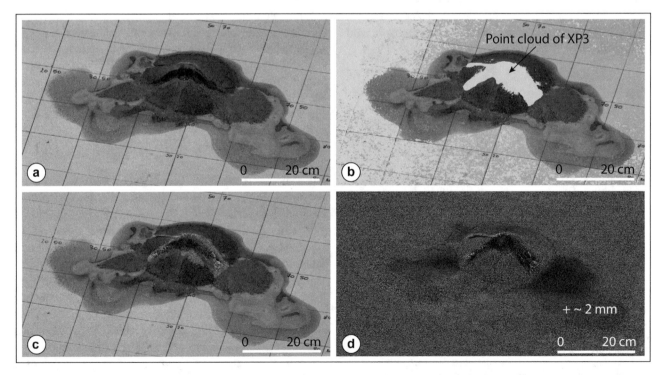

Fig. 3.4 Dense point-cloud of sediment sand-cones' experiments with various water content. **a** Original sand cone after pouring water on the top; **b** original cone overlayed on **a**; **c** modifications around the cone; **d** Imaging of change

3.2 Data Conversion: From Pointclouds to Other Formats

Whether you are using a point-cloud or any form of interpolated dataset, your data remains a discrete sample of a surface, which you can bind (or not) using one method or another. If you think about the accuracy of your dataset to depict the surface you are analysing, you therefore need to think about the error generated by the "choice of feature" represented by the punctual measurement (the process of discretizing the studied object) and if you work with different datasets or if you work with a combination of derivatives (for instance, the slope or the aspect, which for each data point involve a computation relative to other points) from a dataset, you also need to think whether error is generated at this stage as well. Finally, we will remember that the spread of the error will differ depending on the surface itself and depending on the data representation. Imagine that you are working on one of the glacial valleys of Utah in the USA, where you have broad flat-bottom valleys and steep mountain walls, being the remnant of glacial erosion. If you use a gridded dataset that represent the topography on a X, Y grid at 1 m or 0.1 m resolution, the density of measurement per unit surface is going to be very different on the slope than on the "flat": for a sloped surface with angle θ, its relation to the flat surface is given by the cosine of the difference in slope angle (θ) between the two surface, so that for a slope of 60°, with a cosine of 1/2, you are collecting half the points per surface length, in comparison with a flat surface (Fig. 3.5).

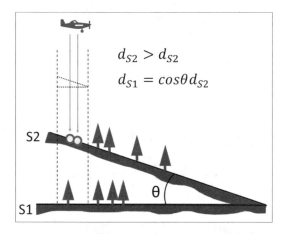

Fig. 3.5 Graphic representation of the role of local slope in modifying the density of collected points per horizontal distance. In this graphic, the two arrows in blue represent two hypothetical laser acquisition of the surface (the points) for a flat surface S1 and a slanted surface S2 (to avoid loading the figure too much, the points on surface S1 are not represented). The slope difference between S1 and S2 is the angle θ. From this graphic, you can clearly see that the surface S2 will get a lower density of points (d) on the figure, than surface 1 (S1) and that this difference is tied by the cosine of the difference of slope angle between the two surfaces

3.2.1 Triangulated Irregular Network (TIN) Data

Whether you are using SfM-MVS or laser technology to collect samples of a surface, it is most likely that one of the steps you will have to take is the meshing of the data, using a TIN. A TIN or triangulated irregular network is made of a set of triangles, built from edges and the points that connect the edges. Because the TIN is made of these three essential components, all three can structure the dataset, and you can have node-based data structure, triangle-based data structure and edge-based data structure (Table 3.2). The choice of one type of structure over another will not be available to you if you are using a GIS or GUI-based environment, but if you call a library in Python or in R, you will have direct access to the different elements of the structure, and you can choose to work with one or another or a combination of the three features.

The generation of TIN datasets from point-clouds can be done by a wide number of triangulation algorithms, and if the student is interested in those issues, as well as in the creation of smarter data structures, the development of effective processing algorithm is most certainly one area of research that needs further expansion, when I am writing this book.

In geomorphology and in geosciences in general, one schema you want to remember is the Voronoi diagrams and the Delaunay triangulation, which are the most common meshing algorithms. The Mathematician M. G. Voronoi, of the turn of the nineteenth–twentieth century proposes to define locations on a plane as a position close to n points, so that for any pick on a plane, one can define which point is the closest. The Delaunay triangulation is accomplished using either a "static triangulation" or a "dynamic triangulation".

3.2.2 Gridded Data

The simplest format and certainly the most widely used geographic system in Earth Sciences is the gridded format, where a division of "row by columns" representing a step size in the X- and Y-direction is providing a record of the topography (Z) or any other data. Because geographical dataset can extend over huge dimensions in X and Y, which have not stopped increasing as sensors provide increasingly spatially dense dataset and as improvement in computing capacities allow us to interact with larger dataset (without untimely crashing), the data structure is divided into multi-level divisions or to borrow a term from the field of computing, they use nested grids (Fig. 3.6). The structure of the data and the scale relation is in such a "multi-level" grid, often stored using a quadtree, which is a data model of the dataset. The tree structure is a series of layers, with each level providing a zoom on a quadrant.

Table 3.2 Traditional TIN data structures	Node-based data structure	Vertex-based data structure	Triangle-based data structure
	The data is represented by the points with their spatial attributes (x, y, z) and the number of neighbouring nodes (ni) and pointers (di) towards those nodes, which take the term of "spoke", so the point-base structure is of the format P(x, y, z, ni, di) for each point	Vertex-based data structure or edge-based data structure are based on the idea that the edges are sufficient to define the triangles, and the neighbours' relations between the points only then need the edges. One traditional structure is the twin-edge structure for instance	The triangle-based structure is maybe the most intuitive, because it includes a table with the points, one with the edges and one with the triangle themselves, which are defined by pointers to the edges forming the triangle. This structure requires more "space" as it stores more variables

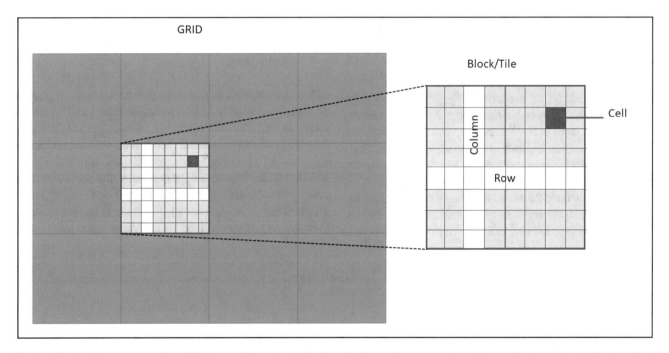

Fig. 3.6 Typical data structure of a gridded data, with a grid, divided into blocks or tiles, or sometimes elongated strips, and each block or tiles is made of series of columns and rows divided into cells of the same size

3.2.3 Making Measurements from a Structured Point-Cloud

In GIS environment from which geomorphologists are used to work from, topographic data is most often in raster format, i.e. an equally spaced grid of information, or in a vector format that can be either a set of RTK GNSS data for instance (a few hundreds to a few thousands for several days' work). In the case of raster data, each gridded data like the altitude, we can name Z on a rectangular coded i, j will be compared to its neighbours to extract the North–South slope for instance (considering that the North–South direction follows the j-direction of the grid:

$$S_{i,j|j} = \frac{Z_{i,j+1} - Z_{i,j-1}}{2n} \tag{3.7}$$

Of course, one can modify the above formula to calculate the slope from only one cell interval, or 5 or 10 or average of group cells, but in one way or another, the user is looking at combining the different cells to extract derivatives. For those who are familiar with differential equations, it is very similar to applying a discretization scheme to solve an equation, where you need to use a grid to generate solutions. For data that is in a vector format, such as a shapefile, the generation and calculation between a two to a set of points require the geomorphologist to interpolate the data to generate topographic data, etc.

When working with pointclouds, the task is slightly different from the one depicted in Fig. 3.6, as we do not have one data by grid cell, but we can have tens to hundreds of them. This is where one of the new challenges for the geomorphologist lies when it comes to data processing.

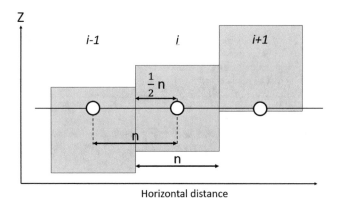

Fig. 3.7 Concepts of binning and structured-pointcloud construction, with a boxing showing all potential values within a single box

Before point-clouds, the geomorphologists had to decide whether the task he or she would engage with was measurable with the topographic data. Do we want to measure pointbar motion in a river from one year to another, for which having sub-metric data is important to volcanic dome growth between the start and the end of a volcanic eruptive phase, for which data at the metre scale may be sufficient. As the data is limited by the scale, it is then easy to discuss the scale issues and the error margin intrinsic to the dataset. When we have a pointcloud that offers a point every couple

of centimetres, the temptation would be to use the all dataset, and getting lost in (eventually) irrelevant micro-variations. Looking back at Fig. 3.7, and assuming that having one data per rectangular cell is the scale at which the data would be most relevant, one would need to weed out the unnecessary data and everything that is not relevant: for instance, removing vegetation from the ground data, etc. This procedure brings back the pointcloud data to a more traditional grid that can be used for computation in a traditional GIS environment.

3.2.4   The Three Most Common Derivatives for Regularly Spaced Pointclouds and Grids

Before we proceed to the main derivatives from a gridded surface, or a regularly spaced pointcloud, let's look at what it means in terms of data spatial structure (Fig. 3.8). Working directly from the pointcloud data has an obvious advantage compared to their gridded counterparts, the calculation is not reduced to the four to height quadrants around a cell (concomitant or along diagonals), but it can be calculated across a potentially infinite array of directions (Fig. 3.8). Consequently, before you decide to transform your point-cloud

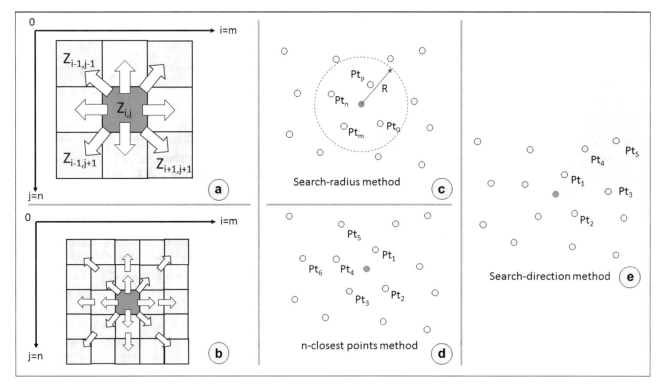

Fig. 3.8 Sample patterns for the calculation of slopes (or other derivatives) from gridded data and from the original point-clouds. **a** Calculation from the neighbouring cells; **b** calculation from the two

nearest neighbouring cells; **c** search radius method around a point; **d** calculation from the *n*-closest points method; **e** calculation along a given direction into a gridded format—as it is the most commonly used format across numerous applications, you are most likely to end with a gridded data—make sure that you are not missing or losing information by gridding the data.

For instance, if you are working on a slope (and you could use a format as in Fig. 3.8 A or B) make sure that the variability of altitude for all the points within one cell is not excessively high and that the statistical descriptor you are going to use does not hide some of the variability in the dataset. One way to do so is to take a few samples of your pointcloud and compare it to the results obtained from the gridded counterpart.

Once you are confident that the transformation into a gridded object is not hiding important information about your point-cloud, you can then use either a properly gridded data or simply a set of points that you have binned into regular voxels or that are regularly spaced along the X, Y plane from which you can use then generate the needed topographic derivatives.

3.2.4.1 The Slope

The first derivative of the topography in both the x-direction and y-direction provides the common derivative of the local slope S in the x- and the y-direction:

$$S_{x,z} = \frac{\partial z}{\partial x} \tag{3.8}$$

$$S_{y,z} = \frac{\partial z}{\partial y} \tag{3.9}$$

As this only provides the local slope, which is highly dependent on the space-step that is chosen, this approach can then be further extended, by multiplying the step size and comparing different scales. If we only look at the topographic change in the x-direction for a time step of 2 cells (you can choose many more and combine different scales to find different elements, then:

$$S_{2x,z} = \frac{\partial z}{\partial 2x} \tag{3.10}$$

And along a discretized form, where the cell where the calculation is performed at cell i, then you can choose to perform the calculation using a space-backward method (Eq. 3.11), space-forward method (Eq. 3.12) or a space-centred method (Eq. 3.13):

$$S_{3(i)} = \frac{S_{3(i)} - S_{3(-2i)}}{3x} \tag{3.11}$$

$$S_{3(i)} = \frac{S_{3(i)} - S_{3(i+2)}}{3x} \tag{3.12}$$

$$S_{3(i)} = \frac{S_{3(i-1)} - S_{3(i+1)}}{3x} \tag{3.13}$$

The choice of the space-step will depend on your dataset and the size of the grid that you have chosen. A purely objective method would compel you to use multiple steps, compare the results at different steps and choose the dataset that creates the best fits at a set of scales. You can also decide subjectively of the right space-step based on your own geomorphological expertise or observations and other measurements in the field for instance. I am convinced that there will be different relevant scales depending on what kind of products you want to create. For instance, you may want to use your dataset with a model, like a weather model, in which case the scale at which you will work will be partly controlled by the space scale of the model.

3.2.4.2 The Curvature

The second derivative of the topography provides the curvature, and using the equations written by [10], the curvature can be written as follows along the x-axis (Eq. 3.14) and along the diagonals (Eq. 3.15), representing the change in both x- and y-direction (i represents the grid horizontal and j the vertical):

$$z_{xx} = \frac{\partial^2 z}{\partial x^2} \approx \frac{z_{i-1} - 2z_i + z_{i+1}}{h^2} \tag{3.14}$$

$$z_{xy} = \frac{\partial^2 z}{\partial x \partial y} \approx \frac{-z_{i-1,j-1} + z_{1-1,j+1} - z_{i+1,j+1} + z_{i+1,j-1}}{h^2} \tag{3.15}$$

3.2.4.3 Local Topographic Variability and Roughness from Standard Deviation

The local topographic variability is a measure of how much variability there is in z in all the neighbouring cells for a given location i. It can be measured using the standard deviation σ, so that:

$$\sigma = \sqrt{\frac{\sum_{i=1}^{n} (z_i - z^2)^2}{n-1}}, \tag{3.16}$$

where n is the population sampled in the measure, and it can be the four connected left, right cells, include the diagonals (eight cells), and even include another row outside this first inner-connected square.

Another potential use of this equation is to investigate the variability of the topographic information within one cell. For instance, if you are using a topographic LiDAR dataset or SfM dataset, you may be able to create several hundred points to several thousand per square metres, but you may work with a grid at a 10 cm, 50 cm or 1 m resolution, for which you will have made a choice of what value to use as a representative topographic value. It might be the maximum, minimum or median or mean value, or you may have generated a density function from which you sample the average upon a statistical parameter of your choice. In such a case, you will have one topographic value and a lot of variability, and this process will exist for every cell of your gridded data. In such a case, you can use Eq. 3.16 to calculate the spatial distribution of variability. For the sake of an example, imagine that you have captured with a mid-range terrestrial laser scanner from multiple vantage points a debris-flow fan (or apron) composed of a large number of blocks and other coarse grains. Your purpose is to run a simple flow routine to see where the debris-flow path was. For this purpose, you decide that a gridded dataset with horizontal resolution of 1 m × 1 m is best, because you can extract the main geomorphology without having a huge number of counter slopes and variations due to the large blocks paving the ground, which would hamper your flow-routing. At the same time, you want to know whether cells looking "similar" at the metre level are including blocks or not, or in other word what is the local topographic variability, below the 1 m scale. For this purpose, you would use Eq. 3.16 to assess the `in-cell' variability. Imagine now that your 1 m gridded dataset displays a channel on the debris-flow fan, but you want to refine the position of your channel banks. One way to do so is to use the same equation once again and find, within each 1 m grid-cell, where do you have major vertical variation, being the result of the channel walls. This problem may be particularly acute if the gridded data was constructed from the minimum z for each cell. It means that for a cell of 1 m × 1 m, even if the cell includes a broad portion of the channel bank, you will only record the lowest value; i.e. you will see this cell as being the channel only.

The standard deviation can be used as a measure of surface roughness, but also to refine your geomorphological understanding of the dataset. Furthermore, there are plenty of other indicators that can be used and that have been listed and clearly explained in [22].

3.2.5 Manual Control of Object Elevation from Dense Point-Cloud with RGB Values

When you create a point-cloud using SfM-MVS or photogrammetry, or when you record the visible spectrum with a camera coupled with the laser scanner (ground or airborne), you can measure the shadow of the object as a control of the vertical height given by the point-cloud (providing that you find sub-horizontal surfaces). Theoretically, this method should not show any discrepancy with a point-cloud generated by more advanced technologies, if not under thick canopy, or in areas where you don't have shadows. The use of shadows is a traditional method in photogrammetry, which relates the length of a shadow L to the height of the object h, so that the angle of the sunray that pass by the top of the object of height h and that reaches the ground at a distance L from the object forms an angle θ, which is related by basic trigonometry so that:

$$\tan(\theta) = \frac{h}{L} \qquad (3.17)$$

To do an accurate measurement, you will need the exact time at which you took the photographs. If you are using a drone or a hand-held camera or a laser with a camera, the data will be time-stamped and you can precisely determine the time of exposure to sunlight. Then, you will need to correct for the Earth declination and the latitude. Once you have those data, you can then compute the anglescanner. The easiest way is to look for the ephemeris table for a given location and date from your favourite search engine online. Traditionally, researchers used an ephemeris table, but you can now have several online calculators and Websites that offer this information freely.

Once you have calculated h from L (i.e. $h = \tan\theta * L$), you can compare the data you are creating with the height given by the point-cloud data. It is a good way to control and make corrections in your dataset, when you have limited targets on the ground for GCPs or when you have patches that are more vegetated than others so that you check the effect of the vegetation on the data quality using this tool. You, however, need to be careful, as Eq. 3.17 is provided for the simplest case when the ground is perfectly flat. You will need to correct the length of the shadow using the local slope or the local topography between the base of your object and the tip of the shadow on the ground, once again using trigonometry if the ground is not perfectly flat.

3.3 Example to Extract Data on a Grid from a Pointcloud with R and LidR

In the R environment, the LidR library offers a function to grid data and make calculations from all the points within this square, dividing the dataset into a set of cells on the horizontal X, Y plane. The LidR library also interacts natively with all the basic descriptive statistical parameters max, min, median, mean, standard deviation, kurtosis, entropy and the different quantiles.

```
library(raster)
library(sp)
library(lidR)
library(ggplot2)

setwd("C:/Users/kaiki/Desktop/")    # set the working
directory where the data is located
las <- readLAS("Estuary2007.las")   # open one LiDAR
file that is used in the chapter on coastal
                                     # geomorphology
```

Calculate the mean altitude for every grid cell of size 50 m × 50 m. You can choose other values. For the figures below, I used 10 and 50 m cell sizes. The different scales will reveal different patterns, and you will be able to extract different types of data

```
elevation <- grid_metrics(las, ~ mean(Z), 50)
```

```
MaxElev <- grid_metrics(las, ~ max(Z), 10)  # same
than above but returning the maximum value
```

```
MinElev <- grid_metrics(las, ~ min(Z), 20)  # same
than above but returning the minimum value
```

```
SdevElev <- grid_metrics(las, ~ sd(Z), 30) # same than
above but returning the standard deviation
```

After generating those different grided set of values, you can work on them as you would do with a raster, and you can also plot them in R using the plot function (Fig. 3.9)

creating your own colour palette, with a bias of the data towards the high values or the low values:

This scale will concentrate the colours around the high values, (Fig. 3.9e)

```
myscale <-colorRampPalette(c("black","blue","-
cyan","pink","white","red"),bias = 0.5)
```

This scale will concentrate the colours around the low values, (Fig. 3.9f)

```
myscale <-colorRampPalette(c("black","blue","-
cyan","pink","white","red"),bias = 0.5)
plot(elevation, col = myscale(100))
```

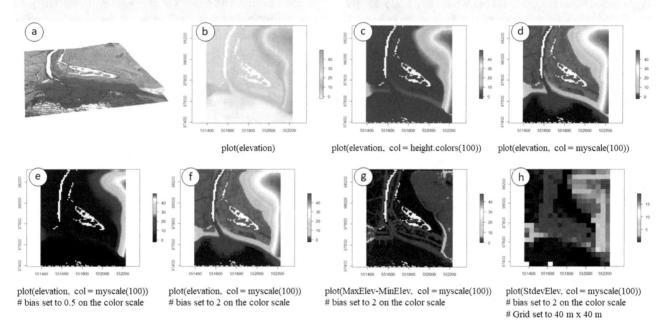

plot(elevation) plot(elevation, col = height.colors(100)) plot(elevation, col = myscale(100))

plot(elevation, col = myscale(100))   plot(elevation, col = myscale(100))   plot(MaxElev-MinElev, col = myscale(100))   plot(StdevElev, col = myscale(100))
# bias set to 0.5 on the color scale   # bias set to 2 on the color scale   # bias set to 2 on the color scale   # bias set to 2 on the color scale
                                                                                                                          # Grid set to 40 m x 40 m

Fig. 3.9 Gridded Lidar data of Cuckmere Haven between the city of Eastbourne and Seaford from data acquired in 2007. **a** 3D panoramic view of the original Lidar of the estuary with the points divided based on the intensity (the sea in red and vegetated slopes orange and yellow); **b** plot of the mean elevation of each grid cell of 10 m × 10 m, with the automatic colour scheme in R; **c** same than **b**, but with the colour set to height, colors divided into 100 units; **d** same dataset, but this time with a custom colour scale (see in the box above how to custom the colour scale), **e** same data with a bias of 0.5, **f** same data with a bias of 2; **g** plot the range of elevation within each grid cell by subtracting the maximum and the minimum of each cell; **h** calculation of the standard deviation, but this time for each grid cell of 40 m × 40 m

3.4 Quality and Error Analysis

In 1974, Prof. James P. Scherz wrote that "It is important that one understand the errors, their sources, characteristics, and relative magnitudes in order to apply photogrammetric materials effectively". Errors in data obtained from photogrammetric methods can be divided between the error due to the limitation of the pixel of the images (i.e. the quality of the sensor) and the photogrammetric process itself. Error analysis is not essential for those interested in the "photogrammology" approach, where a 3D dataset is necessary for visualizing data or obtaining a dataset that is not further subjected to measurement. For all the other usages, error analysis is an important step, as it can be found in all the extracted values. For instance, Fleming and Pavlis [9] have extracted strikes and dips from outcrops, with strike angle differences between 2.6 and 177° and dip differences between 2.5 and 39.8°, comparing field data (which may not be free of error themselves) and SfM-MVS derived measurement using a Nikon and a Sony camera. In their study, there is also a clear difference between the measurement from the Sony camera (Sony DSC-SX9V) generated SfM-MVS model and the Nikon-based (Nikon D5300 DSLD with a fixed falc-35 mm lens) SfM model. Indeed, the strike difference for instance is between 2.6 and 20° with the Sony camera and 8–177° with the Nikon camera (please note that this is not a camera manufacturer issue, but rather a model and lens issue). This is notably why James et al. [14] also required that the camera model and details were communicated in the SfM-MVS protocol they suggested.

As scientists, engineers and researchers in human sciences are increasingly relying on data issues from multiple sensors and multiple measuring methods, the field of error analysis and uncertainty analysis has grown to become an entity on its own. The analysis of uncertainty in geomorphology is an essential step to eventually improve the experiment or the field data collection and processing. It is only after doing a comprehensive error analysis that you will know how representative your findings are. Although error analysis has not always been a fully visible process in quantitative geomorphology, mostly because of its original descriptive and sometimes subjective nature, this tradition has swiftly shifted, notably under the increasing number of engineers doing geomorphology. For point-cloud technologies, there are geomorphologists using photogrammetry and laser scanners, while there are also engineers applying their know-how and methods to geomorphology. This second type of researcher has been particularly instrumental in introducing a more "rigid" (precise) framework in the field of geomorphology.

Depending on the country, this trend has had multiple roots. German geomorphologists have turned towards applied mathematics and statistics in their geomorphologic analysis,

compared to their French counterparts, if not for a few laboratories of the CNRS and Caen and Strasbourg University who have merged slope engineering and geomorphology. In the UK, it is the impulse of a few individuals (centred around Cambridge University originally, and who have spread internationally) who approached geomorphology as engineers. In central Europe, Austria and Poland have also been instrumental in this development and they are rather unique as they have merged a more traditional conceptual framework with new tools and methods of point-cloud acquisition and processing. Finally, in the USA, geomorphology is often located in civil engineering departments, while in Japan contemporary sciences only entered the country at the end of the nineteenth century, with engineering goals, so that the numerical aspect and engineering aspect of geomorphology have strongly prevailed. I would further argue that the University entrance system in Japan, with a strong emphasis on mathematics has also provided the foundation for this new development. Furthermore, strong actors have acted as hinges to redirect the geomorphological sciences, mainly under the influence of Professor T. Oguchi at Tokyo University, who with his students, now turned academics, have favoured this transformation. There are other examples of transformation, but it is invariably the application of methods and tools originating from other disciplines that have come and cross-fertilized the field of geomorphology.

With this numerical evolution, the concepts of error analysis and accuracy estimation have also crossed from physical and engineering sciences to geomorphology.

For the readers interested in developing a complete understanding analysis, I suggest they complete the rapid introduction underneath with other manuals and books that specifically deal with the issue [11, 13]. The themes that are included in the following paragraphs can be found in more detail in these two books.

3.4.1 The Different Types of Error

When you create a point-cloud, you are going to encounter three different types of error, the random error, the systematic error and the mistakes in measurement. In the present case, we are not looking at mistakes and assume that you took your data without any gross error due to negligence or wrong usage of your instrument.

Error handling is an inherent part of any surveying technique, whether you deal with pointclouds or not, and the error in an observation is its distance to the true value. There are different types of errors, random errors and systematic errors (or biases or cumulative errors).

The systematic error is inherent to the instrument and the environment in which the measurements are made. If the

Table 3.3 Three main types of error in the point-cloud data due to position swath angle and range, as well as their effects in 3D and the types of origin or control to look for

Error and noise in the data due to	Origin of the error	Affects
Position	Independent of instruments	X-, Y- and Z-direction
Angle	Altitude and beam angle	Predominantly X and Y
Range	Beam angle and topography	Predominantly Z

instrument remains the same and the environment does not change, the bias is supposed to be stable, but if you start measuring in a foggy environment, and suddenly the weather changes for a bright dry sunny day, it is most likely that your bias is not stable over the all measurement, and this is certainly something you may want to avoid.

Then, the random errors are those that remain once you are sure you haven't made any gross mistake and that you know the bias of your set up. They are statistical errors, and they tend to cancel themselves once you multiply measurements. There are various methods to handle such errors, through interpolation, least square methods, etc.

In the present text, I am concentrating on errors related to pointclouds and won't be dwelling through error handling in surveying (there are plenty of texts on the subject, although most of the surveying error handling can often be applied to pointclouds).

3.4.1.1 Source of Point-Cloud Error

Because this monograph deals with laser-generated and SfM-MVS-generated point-clouds, we will focus on the two specific types of error. These errors can be categorized into different groups beyond laser and photogrammetric technologies:

– As geomorphologists work mostly outdoors, the atmospheric conditions are going to influence both laser propagation and photographs. This influence will come from atmospheric particles, aerosols, rainfall, fog, change in air pressure and the illumination conditions of the scene. For SfM-MVS notably, rapidly moving clouds during an otherwise bright day can be problematic as it will move and create high-contrast shadows (which you should modulate by taking raw format photographs to then work on the shadows).

– Another element of the environment that is affecting both the laser scanner and the photogrammetric method is the reflectance of the surface, and the impact will be different, i.e. a good surface for the laser does not mean a good surface for photogrammetry. For instance, a laser scanner will not particularly like rough surfaces, which can scatter the laser beam, while a photogrammetric method will have more difficulties with a slick dark and wet surface, which will create environmental light mirror effects.

– Finally, the combination between the tool you will be using and the distance and size of the object will be a source of error.

Now, let's turn towards instrument-related error, for both LiDAR (airborne and terrestrial) and SfM-MVS photogrammetry.

3.4.1.2 Error from LiDAR Data

For ALS, the flying height and direction, the laser beam range bias and angular bias as well as the angle of the laser to the surface (i.e. the surface topography) and the size of the footprint, and the Inertial Measurement Unit (IMU), and the GNSS accuracy will impact the quality of the point-cloud. The combination of those parameters will generate two types of error in the dataset, a random noise and a systematic bias (which is not a constant but is also a function of some parameters, like for instance the scattering of laser beam with different weather conditions will not be a constant but a function of both the distance and the "air" characteristics for each location along the laser path). Furthermore, the random error that is often inherent to the measurement technique or the tool itself (imagine it as a loose screw) is actually more complex when it comes to LiDAR and the distribution of the error will generate random but not statistically random error in the dataset (in other word, working on the variance and the covariance as you may do in other cases will not be enough to eliminate the error). Typically, this type of error will be more significant on the outside of the laser strip, and this error will be strongly linked to the topography, the type of land cover and also the altitude of the flight in relation to the swath angle. Error control is also performed in the field using large ground targets with a GNSS at its centre to record what is supposed to be a true X, Y, Z dataset, to which the LiDAR data can then be adjusted or compared for error analysis. Although most geomorphologists will need to understand those errors and where they come from, the dataset that you will use as a geomorphologist is most likely to have been created for you and the point-cloud will be a compound of the errors summarized in Table 3.3. Numerous examples of data processing and error analysis due to LiDAR strip overlapping [12], or raw error analysis at ground control points [7] can be found in the literature and

Table 3.4 Volumetric error distribution for three different SfM software reported by Catalucci et al. [5] comparing a curved bench on a one mast-yacht

Software being tested	Mean (cubic millimetre)	Standard deviation
Agisoft photoscan	5.6638	19.5801
Visual SfM	7.8665	10.4855
Autodesk remake	54.5649	223.882

be used as guidelines for the geomorphologist who would like to go further in this direction.

3.4.1.3 SfM-MVS-Related Error

As a geomorphologist, and a student of geomorphology, you may not have access to a terrestrial laser scanner and a lot of the close-range work you will do will involve SfM-MVS. As an introduction, I advise you to have a look at a recent review of error in photogrammetry, which can be found in [8].

Radial Distortion

The most striking effect of radial distortion is certainly the fish-eye option that most of us have tried on our smartphone camera, where a disproportionate bulge in the centre of the image appears. This fish-eye effect, before being a feature application, is the result of a lens with a strong bulge, so that the projection of the external world on the sensor, or the image plate shows a distortion. This effect exists for most lenses on consumer cameras or on the small UAVs we use for SfM. This distortion can be modelled (and eliminated) using a set of correction equations [20] in i and j on the photograph:

$$\widehat{i} = i + L(r)i \tag{3.18}$$

$$\widehat{j} = j + L(r)j, \tag{3.19}$$

and where $r^2 = i^2 + j^2$ and where L is a distortion function that is calculated from the camera parameters.

Systematic error due to camera factors: Error propagation with SfM-MVS and in photogrammetry is due to the fact that all the rays of light have to pass through the camera length (that has a given thickness, which can eventually be slightly uneven), and be projected on the "photograph" itself. In the simplified case of a pinhole camera, it can be simplified into a hourglass shape projection sketch with the real feature at the bottom and the image of it at the top.

Error from the Software

Volumetric SfM-processing error by software.

Accuracy Estimation

There are two ways to estimate the accuracy of a point in a point-cloud, (1) in the case you have one set that you can consider as being "accurate" and another point-cloud for which you want to estimate the closeness to the accurate dataset, and (2) there is also the case when you don't have a reference dataset to compare your data to.

The Mean of Weighted Observations

With a point-cloud, some of the points might be generated on some tall grass, other bare soil, or concrete, and all of them are combined in a complex spatial pattern in your 3D scene.

The mean of weighted observations returns the mean value of a set of weighted observations of a dataset, for a set of observed points p_1, p_2, p_3, p_4... with the weights being w_1, w_2, w_3, w_4... for each respective observation. If we assume that the weight is inversely proportional to the variance, then we can formulate the standard deviation of the observed data as:

$$\sigma_i = \frac{1}{\sqrt{\sum_i^n w_i}} \tag{3.20}$$

3.4.2 Measuring and Communicating the Error of a Pointcloud

There are several error estimators that can be drawn from a pointcloud, out of which the most popular is certainly the RMSE, but as James et al. [14] wrote it should be avoided or I would argue it should be combined with other indicators as on its own it might not be sufficient to translate the error. Furthermore, systematic error from the measurement technique and random error should also be separated whenever possible. Brasington et al. [2], for instance, have combined the mean error (ME), the root mean square error (RMSE), the mean absolute error (MAE) and the standard deviation error (SDE). Finally, please note that the software you are using may also provide you some error measurements, due to the process itself. The SfM-MVS software *Photoscan Pro* and *Metashape Pro,* for instance, both provide you with error estimations of the SfM and registration process. In Agisoft, it is possible to place a-posteriori quality control points, so that for a known expected value one can compare it with the calculated value as well as the error due to the registration process as provided by the software. For instance, Marteau et al. [16] implemented a bootstrapping resampling method on the ground control points to generate quality control points or check points, from which they calculated the residual error. They used the standard deviation of these residuals to state the error and the mean of these residuals as the accuracy level.

In this section, I am providing you with the major error indicators (it is not an exhaustive list, and you can also create some of your own), and then some simple suggestions on how to implement those easily in R using the R-studio environment.

Root Mean Square Error (RMSE)

The RMSE is certainly the most popular method to account for errors in sciences and engineering. However, it has to be handled carefully, because it does not provide a direction to the error, and in a space with regularly spaced "bumps", like a gravel surface for instance, SfM-MVS may tend to ignore the deepest holes in between gravels, because they are in the shadow, and the highest points of each gravel may also be exaggerated in Z (the vertical stretch for mountains and other "pointy" object is often a difficulty with SfM-MVS), in which case the RMSE might be low. In this case, it is ignoring both extremes and may provide a biased view of the error. However, in most cases, the RMSE still provides you with a good estimate of the error.

$$\text{RMSE} = \sum_{i=1}^{n} \sqrt{\frac{\left(\widehat{X}_i - X_i\right)^2}{n}} \qquad (3.21)$$

Mean Bias Error (MBE)

Because of the limitations of RMSE, combining RMSE with MBE is an interesting combination, as it does not square nor turn the values into absolute values, this indicator will give you an indicator of the main directionality of your dataset error.

$$\text{MBE} = \frac{1}{n} \sum_{i=1}^{n} X_i - \underline{X}_i \qquad (3.22)$$

Mean Absolute Error (MAE)

The MAE provides a descriptor for the average magnitude of the error in a set of predictions against the expected values. It differs from the mean bias error (MBE), as it does not

consider the "direction" of the error, as it uses the absolute value of the difference between predicted and expected values in a population.

$$\text{MAE} = \frac{1}{n} \sum_{i=1}^{n} |X_i - \underline{X}_i| \qquad (3.23)$$

Although the MAE and the RMSE look rather similar in their formulation, the two have different behaviour depending on the shape of the distribution and the frequency of outliers. The RMSE value will tend to increase with the increasing relative-importance of "extremes" and outliers, while the MAE will keep a lower value. As an indicator of the frequency of outliers and their influence on the distribution, one may want to compute both rather one over another.

Standard Deviation of Error (SDE)

$$\text{SD} = \sqrt{\frac{\sum_{i=1}^{n} |X_i - \underline{X}|^2}{n}} \qquad (3.24)$$

Standard Error (SE)

The standard error is the variance divided by the root of the population number

$$\text{SE} = \frac{\sum_{i=1}^{n} (X_i - \underline{X})/(n-1)}{\sqrt{n}} \qquad (3.25)$$

Obtaining Some of the Most Common Error Parameters in R

If you haven't already done so, you can download the R and R-studio free version to your favourite environment (available on Windows, OSX and Linux), and this may be the most time-consuming part, all the rest is very simple.

First, you need to set up your work directory, either in R or in R-studio; in the present case, my data are in the D drive in the folder Experiment 3, and from this location, I read the sparse pointcloud that is saved as data .txt and that is loaded in *R* as the "sparse" data:

```
setwd("D:/LabExperimentsUC-floor/Experiment3")

sparse <-read.table('data.txt')
```

You could write your own function in R, but the beauty of this environment is the availability of libraries to do almost anything, and we will use the library "Metrics". If you have already installed the package, you can skip the first line:

Install.packages(Metrics)

Library(Metrics)

From this library, you can obtain the root mean square error, the mean absolute error and the median absolute error for instance as follows (for further error functions, please see the pdf on the CRAN Website: https://cran.r-project.org/web/packages/Metrics/Metrics.pdf.

In the example underneath, we created a vector of expected values, which are all $Z = 0$ altitude for our laboratory experiment:

```
sparseExp <- numeric(nrow(sparse))

x <- as.numeric(unlist(sparse[3]) # the Z values are
stored in the third column

rmse(x, sparseExp)  # calculate RMSE, you could
add "variable <- " to assign it

                    # to a variable

mae(x, sparseExp)   # calculate the MAE

mdae(x, sparseExp)  # calculate the MDAE
```

3.4.3 Error Propagation

Finally, to conclude this short introduction to error and error analysis, you will also have to deal with error when you are multiplying two datasets, or when you are adding them, and when you multiply two datasets, you also need to know how to "multiply" the error. Similarly, if you transform your dataset, or you add it to another dataset, you need to understand how to handle the error, and this is also part of the process of "error propagation". The error propagation is then generally expressed as a function of the point being recorded, and this function is not dependent of the other values (in general theory, but there are special cases when this assumption is not valid), so that you can express error as (Eq. 3.26):

$$E_{fp} = \sqrt{\sum_{i=1}^{n}\left(\frac{\partial f}{\partial i}E_i\right)^2},\qquad(3.26)$$

where

E_{fp} is the error propagated for a pointcloud p by the function f,
E_i is the error at a point i in the population 1 to n.

In such a way that the error is related to the partial derivative of the applied function at each point.

Let us say for instance, if you are adding measurement with error, and that you choose your error value to be one of the sigma of the distribution of your data, then the error propagation becomes (Eq. 3.27):

$$E_{\text{sigma}.p} = \sqrt{\sum_{i=1}^{n}(\sigma_i)^2}\qquad(3.27)$$

Similar error manipulations can be used to multiply, divide, etc., your point data.

Once you are at this stage and you have the different indicators of the error from your pointcloud, you are ready to use it for geomorphological applications. You know what are the limits of your pointcloud (and hence what you can say with it). Remember, it is not because you have a pointcloud with for instance a metre of vertical error that you have to throw it away. If you are working on mountain-scale features in the Himalaya or the Andes, nobody cares about a metre error. But, if you are planning to do micro-topography, or soil erosion measurements over a few years with repeat survey, in such case you want to minimize the error as much as possible. And this last consideration brings us back to Chap. 2 and the importance of having a lot of ground control points, well spread over the surface you are interested in, and also the need to test your data and eventually go back to the field where you collected the data, to then add more control points, modify your sampling strategy. It is therefore not a one-shot process in most cases, but rather a trial and error process, in which the error calculation process will be essential in telling you whether you should and could proceed with the data that you have collected. This is one of the main danger of pointcloud technology: it looks good even wrong data looks good. Especially SfM-MVS will create some pointcloud in most cases, but it can also generate weird shapes, as it is never just a geometric and statistical approach to the geometry.

Conclusion
At the end of this chapter, you should now know what are the different steps to transform your original raw pointcloud into a registered pointcloud that you have then filtered to obtain (as a geomorphologist most likely) the ground points, and you should be able to provide indicators and measures of the error inherent to your point-cloud: the error due to the machine, due to the point-cloud processing and data transformation.

This chapter is the last one of the introduction section on point-cloud acquisition and processing, and in the remaining chapters, we will look at various applications found in the literature and also go through a few examples;

notably reusing the LidR package, we have briefly encountered in these chapters. We will have examples of data filtering and processing, so that the steps we have seen together in the first chapters can be understood in context.

Potential Projects to Learn by Doing

(1) Download a small pointcloud dataset from Opentopography.org or use any of your own dataset and then grid the data, so that you have a grid over your dataset of (let's say) 50 × 50 quadrangle. Repeat the same operation with 40 × 40, then 30 × 30 [...] to 5 × 5. For each value taken in a quadrangle, calculate MAE, MBE and RMSE to the original pointcloud. Determine the role of the gridding process and the role of the grid-step on your results. Plot the result on a graph with the grid size and the different errors. Now, what is the most effective scale for you to work? What scale would you use for what kind of work (i.e. imagine different potential industrial and research-related usages of your gridded data. Reflect on which scale would be the most appropriate)?

(2) Create a pointcloud with SfM-MVS of a surface (let say a table covered in sand or anything as such), then add one protruding object like a sand cone in the middle. Now, compare both pointclouds (with and without the sand cone). Is the error spread the same way? Do you see distortion based on the object shape and their surroundings?

References Used in the Chapter and Further Reading

1. Asal FFF (2019) Comparative analysis of the digital terrain models extracted from airborne LiDAR point clouds using different filtering approaches in residential landscapes. Adv Remote Sens 8:51–75
2. Brasington J, Langham J, Rumsby B (2003) Methodological sensitivity of morphometric estimates of coarse fluvial sediment transport. Geomorphol 53:299–316.
3. Brodu N, Lague D (2011) 3D terrestrial lidar data classification of complex natural scenes using a multi-scale dimensionality criterion: applications in geomorphology. ISPRS J Photogramm Remote Sens 68:121–134
4. Butler H, Chambers B, Hartzell P, Glennie C (2021) PDAL: an open source library for the processing and analysis of point clouds. Comput Geosci (in press)
5. Catalucci S, Marsilli R, Moretti M, Rossi G (2018) Comparison between point cloud processing techniques. Measurement 127:221–226
6. Chen Z, Gao B, Devereux B (2017) State of the art: DTM generation using airborne LIDAR data. Sensors 17:150. https://doi.org/10.3390/s17010150
7. Csanyi N, Toth C (2007) Improvement of LiDAR data accuracy using LiDAR-specific ground targets. Photogramm Eng Remote Sens 73:385–396
8. Dai F, Feng Y, Hough R (2014) Photogrammetric error sources and impacts on modelling and surveying in construction engineering applications. Visual Eng 2:1–14
9. Fleming Z, Pavlis TL (2018) An orientation based correction method for SfM-MVS point clouds–Implications for field geology. J Struct Geol 113:76–89
10. Gallant JC, Wilson JP (2000) Primary topographic attributes. In: Wilson JP, Gallant JC (eds) Terrain analysis: principles and applications. Wiley, Hoboken, pp 51–85
11. Gupta SV (2012) Measurement uncertainties—physical parameters and calibration of instruments. Springer, Berlin, 321p
12. Habib A (2018) Accuracy, quality assurance, and quality control of light detection and ranging mapping (Chapter 9). In: Shan J, Toth CK (eds) Topographic laser ranging and scanning, principles and processing, 2nd edn. Routledge, Milton Park, pp 291–312
13. Hughes IG, Hase TPA (2010) Measurement and their uncertainties—a practical guide to modern error analysis. Oxford University Press, Oxford, 136p
14. James MR, Chandler JH, Eltner A, Fraser C, Miller PE, Mills JP Noble T, Robson S, Lane SN (2019). Guidelines on the use of structure-from-motion photogrammetry in geomorphic research. Earth Surf Process Landf 44:2081–2084
15. Kraus K, Pfeifer N (1998) Determination of terrain models in wooded areas with airborne laser scanner data. ISPRS JPRS 53:193–203
16. Marteau B, Vericat D, Gibbins C, Batalla RJ, Green DR (2017) Application of structure-from-motion photogrammetry to river restoration. Earth Surf Process Land 42:503–515
17. Pinton D, Canestrelli A, Wilkinson B, Ifju P, Ortega A (2020) A new algorithm for estimating ground elevation and vegetation characteristics in coastal salt marshes from high-resolution UAV-based LiDAR point clouds. Earth Surf Proc Land 45:3687–3701
18. Rowley T, Ursic M, Konsoer K, Langendoen E, Mutschler M, Sampey J, Pocwiardowski P (2020) Comparison of terrestrial lidar, SfM and MBES resolution and accuracy for geomorphic analysis in physical systems that experience subaerial and subaqueous conditions. Geomorphology 355(107056):1–13
19. Sithole G (2001) Filtering of laser altimetry data using a slope adaptive filter. Int Arch Photogramm Remote Sens XXXIX-3/W4:203–210
20. Tsai R (1987) A versatile camera calibration technique for high-accuracy 3D machine vision methodology using offthe-shelf TV cameras and lenses. IEEE J Robot Autom 3:323-344.
21. Vosselman G (2000) Slope based filtering of laser altimetry data. IAPRS XXXIII-B3:935–942
22. Wilson JP (2018) Environmental applications of digital terrain modelling. Wiley, New York, 336p
23. Zhang K, Chen S-C, Whitman D, Shyu M-L, Yan J, Zhang C (2003) A progressive morphological filter for removing nonground measurements from airborne LIDAR data. IEEE Trans Geosci Remote Sens 41–4:872–882

Point-Cloud Technology for Coastal and Floodplain Geomorphology

4

Abstract

This chapter starts the section of applications of point-cloud technologies in geomorphology, and as this section starts at the shore, going up the mountain. This chapter presents the advantages and difficulties of working and collecting pointclouds in coastal and floodplain areas. The chapter starts with some very basic ideas on river catchments, sediment transport and erosion creating river landscapes, as well as the main processes modifying the coasts. This first section acts as a rapid presentation to non-geomorphologists before presenting the pointcloud technologies and the floodplains, with a space division at different scales showing how different landforms need to be approached with different data acquisition methods. The same division of space by scales and landforms has been then applied to coastal landforms, and advantages and disadvantages of the different methods are also presented. The chapter then ends with two sets of worked examples using a meander of the Alabama River (USA) and a coastal example in Cuckmere Haven in the UK.

Learning Outcomes: Be Able to Apply Your Skills to the Coast and the Floodplain

- Have a basic understanding of some key landforms and processes of coastal and fluvial geomorphology;
- Understand the link between coastal and fluvial landform morphologies and adapted point-cloud technologies;
- Understand the limitations and advantages of the different methods for different landforms;
- Be able to draw profiles and grid data from SfM and LiDAR data;
- Be able to manipulate and transform *.las files to extract other files usable in GIS and extract different derivatives useful in fluvial and coastal geomorphology.

Introduction

Coastal and fluvial geomorphology are challenging environments for 3D pointcloud data acquisition, because they extend over long strips, often narrow, and moreover, they offer vertical variation that is several folds smaller than the horizontal extent. For instance, the Cuvelai basin floodplains in Namibia, although mostly deprived of vegetation—eliminating one difficulty—are only showing vertical variations of a few tens of centimetres over hundreds of square kilometres. The temporary levees often change place "randomly" with the next flood in this archaic environment [89], making the use of HRT pointcloud an essential tool.

Most often than none, coastal environments, rivers and floodplains are synonyms with water and vegetation, signifying that LiDAR technology providing a method to see "through" and in between the vegetation has been highly instrumental in researching these environments. Furthermore, coasts, rivers and floodplains are essential societal and development assets, in such a way that those environments have attracted researchers, funding and practitioners. If you look at a world map and search for the major cities, the major investment areas, where most of the infrastructures are, where most of the agricultural land is located, we will all agree on the importance of precise, if not accurate, measurement of the coasts and floodplains.

Consequently, it will come as no surprise that floodplain and river research has been the main engine of point-cloud technology development in geomorphology. In a recent article (at least recent, when I am writing these lines), two of the (arguably) world leaders in fluvial geomorphology—Prof. Kondolf and Prof. Piegay—and their team have compiled data from over 1700 journal articles, showing the emergence of a "new period" in the field of fluvial geomorphology, starting around year 2000, thanks to the emergence of new technologies [90], i.e. computing technologies and data acquisition technologies. This shift is part of a more general evolution in the field of geomorphology,

© Springer Nature Switzerland AG 2022

C. Gomez. *Point Cloud Technologies for Geomorphologists*.
Springer Textbooks in Earth Sciences, Geography and Environment.
https://doi.org/10.1007/978-3-031-10977-5_4

which has experienced a progressive modification of the way the "natural philosopher" who became scientist (after 1833) has modified its approach to fieldwork, and so mostly under the influence of technological advances [25]. Unsurprisingly, high-resolution topography and point-cloud technologies (both acquisition and processing) are a part of this technological advance. If, like myself when I was a student, you are or were a cheeky or insolent student, you would certainly ask: "and so what, technological advances happen all the time!". Well, I would argue that high-resolution-topography and point-cloud technologies are more than a mere advance, it has been a revolution in what can be scientifically done in fluvial geomorphology, and it has had repercussions in all the contiguous research fields. Indeed, because river and floodplain morphology extends spatially over several hundred metres to tens of kilometres across the floodplain, for often vertical variations that can be only within a couple of metres (and sometimes much mode of course), high-resolution and high-accuracy topographic data acquisition has allowed scientists to measure features that were not visible or measurable before (unless you were ready to spend your life with a geodetic instrument walking the field till you drop). Furthermore, models that could only be built from approximated topographies, or cross sections could suddenly be filled with complete topographic and bathymetric datasets. Potential errors and limitations of an ill-chosen cross section were suddenly not an issue anymore. A single roughness data on a gravel point-bar for instance can now be replaced with a spatial variation of surface roughness calculated from point-clouds, etc. Point-cloud is indeed just a technical advance, but a technical advance that has modified the paradigms and the possibilities of a wide array of scientific research and engineering.

4.1　Introductory Few Words on Coastal and Floodplain Geomorphology

4.1.1　Floodplain and Channel Geomorphology

Because of the importance of rivers in human communities and societies (freshwater access, transport, flood control, irrigation…), river and floodplain geomorphology is of great interest beyond the field of geomorphology. The USGS writes "An understanding of river- and stream-channel geomorphic responses to various human-caused and natural disturbances is important for effective management, conservation and rehabilitation of rivers and streams to accommodate multiple, often conflicting, needs. Channel changes may have implications for the protection of property and structures, water supply, navigation and habitat. The channel-bank erosion that accompanies natural channel migration on a floodplain represents a constant threat to

property and structures located in or near the channel. Various human-caused and natural disturbances include reservoirs, channelization, in-channel sand and gravel extraction, and urbanization. A common natural disturbance is a flood. Possible geomorphic responses of a channel to disturbances include channel-bed degradation (erosion), channel-bed aggradation (deposition of material), channel widening, and channel straightening. These adjustments represent the channel's attempt to establish a new approximate equilibrium condition" [111]. This statement emphasizes the importance and the complexity of the different socio-economic ecosystems and environmental ecosystems around waterways and the influence of geomorphological knowledge beyond the discipline.

Fluvial geomorphology is the study of the interactions between the river processes and their landforms at a broad range of times and scales [23], with rivers flowing continuously over the year (perennials), or only some parts of the year or some years (ephemeral), and they vary depending on whether they are cut in bedrock or whether they flow over alluvial channels. The channel gradient and the sediment deposition and transport then defines different channel planforms (braided, meandering and straight channels). Those forms are also linked to the frequency and proportion of discharge variations over a year and over several years. If one "unzooms" and takes a virtual bird-eyes view of the catchment, then the shape of the drainage network (dendritic, radial, etc.) will appear, showing the role of the underlying structural geology and change in base level on the river network. In other words, there isn't one single fluvial geomorphology, but a wealth of various systems, by which sediments and water are conveyed from high-topographic points (mountains…) to sinks (other rivers, lakes and the ocean). For the readers who are interested on the topic of fluvial geomorphology, I can only suggest the book "Tools in Fluvial Geomorphology" edited by Kondolf and Piegay, which provides amidst tools and methods a broad view on fluvial geomorphology [58].

In this first subsection, I have tried to present the different relevant systems and processes following a logic that can be useful to the geomorphologist who is interested in point-cloud technologies. The student will find information to discuss and explain the morphologies generated from point-cloud technologies. Further reading is obviously advised, but it may give some ideas of the different possibilities of analysis and the linkages that can be made to mathematical and physical modelling or geomorphologic process modelling.

River Catchment and Drainage Organization
Rivers and streams are organized within units named river catchments and drainage basins. Each river catchment and drainage basin are organized around one main stream, to

which the water of tributaries also contribute. Sunamura [107] first defined it as an open system, where energy from the atmosphere, the Sun, with a main water and sediment outlet. This set of interconnected streams within the river catchment creates spatial patterns that are an expression of the geology, the slope of energy, and the atmospheric processes (rainfall patterns, frequency, air temperature, moisture, etc.). The drainage patterns are also useful tools as a first assessment of the geology and geomorphology of an area. They are divided into sets of patterns, the main ones being the dendritic patterns, which is the most common type, with branching tributaries with an angle with the main stem < 90°. Then, there are the radial patterns, that can be found on stratovolcanoes like Mount Merapi in Indonesia, where the subsurface structures are parallel to the slope, the parallel drainage patterns, where slopes are steep, the trellis or lattice pattern, that shows sub-parallel streams with small-size tributaries that reach the main streams almost perpendicularly to it. This qualitative assessment of the planform network of streams can then be further complemented with numerical analysis, such as the Horton's law of streams: consider a catchment with S segments and n is the order of the catchment, we can then then link the log of the number of catchments of a given (S_n), from which the bifurcation ratio (R_b) and a linear regression with two constants $C1$ and $C2$ can be expressed as (Eqs. 4.1 and 4.2):

$$\log(S_n) = C1 * n + C2 \qquad (4.1)$$

and:

$$\frac{S_n}{S_{(n+1)}} = R_b \qquad (4.2)$$

This expression of R_b has been further shown to be in part dependent on the stream order and was rewritten by Schumm [98] as the "weighted bifurcation ratio" WR_b (Eq. 4.3):

$$WR_b = \frac{\sum S_{n(n,n+1)}(S_n + S_{n+1})}{\sum S} \qquad (4.3)$$

The Generation of Streams and Catchment Landforms: Erosion and Sediment Yield

The variety of landforms, their shape, size and rhythms of evolution are all the resultant of erosion and sediment yield (transport and deposition). These processes are particularly important for river geomorphology, because the later are the conveyor-belt of sediments from catchments to the sea. Estimated export rates can reach 28,000 tonnes/km^2/year for the Tsengwen River in Taiwan or 55,500 tonnes/km^2/year

for the Hangfuchuan River in China, for instance, the two rivers having catchment areas of 1000 and 3199 km^2, respectively [114]. In streams and rivers, the sediment will travel either as bed-load rolling on the floor or moved by saltation or as suspended load. Consequently, the size of the grains (gravels), the presence of smooth sandy and sand and silt point-bars, or the bedforms exposed after a flood, which you will measure using SfM-MVS or a laser scanning system, will thus provide you with data not only on the bedform, but also on the hydraulic and the hydrology of the flow that created the landform (either by erosion, deposition or a mix of both).

Furthermore, point-cloud technology can also be used at the local scale to characterize the change of the material size along the river. For rivers flowing along steep gradient, like the ones in the West Coast of the Southern Alps of New Zealand, starting, for instance, from the Fox [91] and the Franz-Josef glaciers [31] in New Zealand, rivers can transport gravels from the mountain down to the sea, and it is therefore possible to use point-cloud technology to monitor and survey the characteristics of those gravels over time for instance, to extract river energy information. Point-cloud technologies are therefore offering the coming generation of fluvial geomorphologists with a new breath of opportunities to understand fluvial environments.

I think that the students of geomorphology who will read (or skim through) this book will need to use their imagination and think about what is now possible to do with this "newish" technology and how you can then improve the work that has been done by your peers.

One well-known empirical approach relating sediment size and discharge uses the critical shear stress. It was proposed by Shields in 1936. According to his analysis, the incipient motion of a sediment particle of size d located on the bed, and based on the Karman–Prandtl's equation of turbulent flow velocity distribution, is Eq. 4.4:

$$\frac{u_d}{u_*} = f\left(\frac{u_* d}{v}\right), \qquad (4.4)$$

where u_d is the velocity just above the particle to be moved, u_* is the shear velocity, d is the particle size and v is the kinematic viscosity. From this approach considering the critical shear stress, researchers have then developed a formula to link the mean velocity of the flow, as it is related to the shear stress on the bed. The critical velocity related to the Shields parameter has thus been investigated by different authors (e.g. Table 4.1):

For the sake of giving another example, we can consider the commonly used Froude number that directly links the bedforms to the flow, and [43] has given those relations (Table 4.2).

Table 4.1 Values of the shields parameter and critical velocity related to the water depth and the sediment sizes (d is the particle diameter, and D is the depth of flow in SI units)

$\dfrac{U_{cr}}{\sqrt{\frac{\gamma_s d}{\rho_f}}}$	References
$1.414\left(\frac{D}{d}\right)^{1/6}$	[77]
$0.5\mathrm{loglog}\left(\frac{D}{d}\right) + 1.63$	[37]

Table 4.2 Relations between the Froude number and the associated main types of bedforms

Froude number (flow characteristics)	Bedform associated
$\ll 1$	Ripples and flat bed
< 1	Dunes
> 1	Antidunes

Those numerical approaches have come mostly from the field of engineering and been used by geomorphologists, but because the level of precision and accuracy of the topographic or bathymetric data that could be acquired was limited before the advent of point-cloud technologies, it was not really possible to refine these relationships further from the perspective of geomorphology, if not for attempts to use planform shapes notably.

Now, with point-cloud technologies, it is possible to use small variations in the shape, the size, vertical change in those bedforms, and link them to grain size and flow rheology. And this work can be done at a multitude of scales. For instance, ripples will vary between 0.5 and 2 mm, dunes between 0.4 to several hundred metres, sand bars a few tens of metres to several kilometres, etc., and point-cloud technologies will be adapted to work across those different scales. You will therefore have to choose the right tool and sampling method to extract the necessary information to not only relate landforms and bedforms to flow characteristics but develop methods that make full use of the HRT point-cloud data to quantify flow characteristics. This scientific development, then in turn, can enhance connected scientific fields, such as sedimentology, where scientists attempt to link quantitative sedimentology to flow data using notably the Anisotropy of Magnetic Susceptibility (or magnetic fabric) from sand grains [116] and for which HRT is an essential dataset.

From this rapid brush through, you can see that—and it is the case in most geomorphological studies—streams, rivers and floodplain geomorphologies are imbrications of different scales, making the development of point-cloud technology a windfall, as it allows the scientists to collect dense samples, with high accuracy and precision, and so over wide areas. It is therefore easy to bridge the different scales, without having to reflect on the necessary adjustments of one dataset with another. The first section then finishes with sets of tools and methods that are used to estimate sediment transport, erosion and deposition in river and fluvial systems, and

I chose to accentuate this section slightly more than in other chapters, because river erosion, deposition is dominated by Newtonian flows (stress and strain are linked by a linear relationship), and because these relations have given rise to mathematical and physical relationships between sediments, landforms and flow characteristics. The same idea stands for most coastal processes, but not for debris flows or lahars on volcanoes or pyroclastic flows. Indeed, because scientists do not understand fully the physical processes that lead to erosion and deposition, and because there is no unified theory (for debris flows or pyroclastic flows, scientists use numerical estimation methods, but they are based on numerous assumptions rather than a complete physical understanding of the processes occurring—which is an exciting news if you are a student, because it also means that there is still plenty of research to be done). As a consequence, for river processes, there is the possibility to obtain numerical values on a flow, based on a combination of precise landform and bedform quantification. And to do so, the best tool is arguably a point-cloud.

4.1.2 Coastal Processes and landforms—A Rapid Glance at the Processes that Make and Change the Coast

"Short-term processes" at the human scale—Out of all the processes acting on the coast, the "big four" that modify and shape a coasts are the coastal current, the tide, the wave, the wind (all interacting with one another) and the more occasional tsunamis and storm surges. At a longer time scale, the relative sea level, which is mostly controlled by the relation of the amount of water in the oceans and the geometry of the latter, has a preponderant role in shaping the coastline.

(a) **Coastal currents**—Coastal currents are the results of winds, regional sea currents and wave characteristics.

For instance, the breaking of waves at the coast creates a backwash underneath as well as an orthogonal swash, forcing water flows and displacing sediments. Water discharge from the rivers, also create river currents accompanied by sediment transport, having direct effects on the coast and on the other coastal currents. In turn, those currents will have very different effects whether they occur against the direction of trains of waves or in the same direction. In the first case, waves

(b) **Tides**—Tides are certainly one of the best example of the interrelations in geosciences, as they are the results of the alignment patterns of the Earth, Moon and the Sun, with the two perigean spring tides being the largest, when the Sun, the Moon and the Earth are aligned in this order and the Moon is at its closest to the Earth. This vertical movement is accompanied with inland progression as the tide rises, and it is also the origin of underwater currents.

(c) **Waves**—Except for waves triggered from the bottom of the ocean by earthquake or landslides, in which case the all column of water above is translated, waves are surface undulations produced by wind action arriving at a 12–16 s intervals from several centimetres to several metres at the shore, while they can reach more than 20 m in height in the open ocean. At the coast, the geomorphic potential of waves is controlled by the height, speed and frequency of the waves, as well as the height change of the ocean, which can swell during spring tides for instance but also during storms, when the atmospheric pressure is lower, and thus, the ocean level is higher. If waves can travel parallel to one another, they usually show more complex planform shapes near the coasts due to refraction processes with the coastal land and the shallow bathymetry.

(d) **Wind effects**—The last process, which this time occurs not in the water but over it and over land, are winds. Wind has an indirect effect on the coasts, by propelling waves and currents, and it also has direct effects on the aerial landforms, such as sand dunes. It also transports the fine-sediments from the beaches and the tidal flats. These sediments are then transported inland to form the coastal dunes for instance.

(e) **Storm surges and tsunamis**—Finally, there are processes that have a long recurrence period, and which can be fathomed as random at the human scale. Storm surges are a combination of wind and low-atmospheric pressure processes, while tsunamis are the results of a sudden and "relatively localized" displacement of a large if not the all column of water from underneath with earthquakes for instance, or when a large body of material slides into the sea (for instance, Hawaii volcanoes collapsing into the sea), or when a meteorite impact the Earth ocean and displace a large volume of water, if not the full depth of the ocean as submarine impact craters suggest. Those waves have in turn "catastrophic" and long-lasting effects at the coast.

Longer-term processes: relative sea-level change—If we fence our argument within the Holocene period, relative sea-level change becomes the main driver of coastal evolution, landforms creation and migration. The term used here is "relative" sea-level change, because changes of the geometry of the landmasses (e.g. subsidence), change in the geometry of the ocean, as well as the distribution of water on the planet can all generate sea-level variations.

4.2 Point-Cloud Technology and the Floodplain

4.2.1 SfM-MVS and LiDAR for Floodplain Geomorphology

SfM-MVS and other HRT methods have been first tried and imported to the field of fluvial geomorphology, before percolating to other fields of geomorphology (there are exception, but if you are looking for the arrow in History, this is it).

Indeed, floodplain and river geomorphologists have to deal with vertical changes in the fluvial and floodplain landscapes that can be a few tens of centimetres only and so over areas of a few hundred metres to kilometres. A good example of this issue are the flood levees of the Cuvelai Basin in Namibia that are not higher than a few tens of centimetres but that extend over several kilometres (e.g. [89]). In such a case, one needs to have a high precision and accurate altitude, and this precision needs to be available over extensive areas. To this first reason, we can add a historical reason. Indeed, river geomorphologists have been working alongside civil engineers for evident reasons, and I would argue that this exposure are cross-fertilized both fields, with geomorphologists using civil engineer tools and civil engineers also applying their skills to the field of geomorphology. These collaborations have further extended with the development of river restoration, looking at the effects of infrastructures and restoration projects on the sediment transport for instance [56] or to the contrary on the impact of rivers on infrastructures (e.g. [115]). The well-known book and manual of techniques and methods in the field of fluvial geomorphology [57, 58] is a very good example where one can see how methods that originated from engineering, statistics, hydraulics, etc., all merge together to provide tools to the fluvial geomorphologist. On top of these mainstream directions, there are local and national specificities that come into play. For instance, in Japan, because contemporary science only started to exist with the Meiji Restoration period at the end of the nineteenth century as the result of the economic need to include the latest engineering developments (road network, bridges, dams, etc.), science is still in many ways the vassal of engineering. It has negative effects on the scientific development of the nation, but it has also allowed Japan to merge fields that were traditionally separated in Europe and to apply engineering numerical methods to geomorphology notably (n.b. this is not to say that there was no scientific tradition in Japan before the Meiji era).

If we leave Japan and go back to the general evolution of the field of fluvial geomorphology, it is therefore within this framework of cross-fertilization with civil engineering (mostly) that numerous of the recent progress in monitoring techniques have been spear-headed in the field of fluvial geomorphology. For SfM-MVS, this is further accentuated by the predominant role that the British school of geomorphologist and civil engineering has taken in this field. To date, the most influential and pioneering work in transferring civil engineering methods to geomorphology can be linked to a group of civil engineers from the UK with predominantly three figures: Prof. Jim Chandler, for his applied work in photogrammetry [16–22], as well as the work of Prof. S. N. Lane who also was a pioneer in the field of photogrammetry applied for fluvial geomorphology [62–64]. And like Prof. Lane, also coming from the Civil Engineering Department of the University of Cambridge, Prof. J. Brasington has also been instrumental in those developments [9, 10] and contributed to some of the most popular work in applied SfM-MVS in geomorphology as a whole [117].

Clearly, it is under the pulse of this group of civil engineers interested in geomorphology that a transfer of knowledge and skills started to operate at the turn of the twentieth–twenty-first century and has led to the spread of SfM-MVS in geomorphology notably, and if those scientists have been "messengers", the kick-start that popularized the method was most probably the work of K. N. Snavely from Internet photo collections [103, 102].

The transfer of LiDAR technology to geomorphological application marks a corner-stone in geomorphology. And more precisely, it is the field of fluvial geomorphology that did first manifest the strongest interest in the method [65]. Indeed, LiDAR high-resolution topography (HRT) is a powerful tool in fluvial geomorphology, because the vertical topographical changes of rivers and floodplains are often limited and spreads over hundreds of metres to kilometres, making LiDAR technology an invaluable asset for morphological analysis, chronological analysis and then to obtain precise boundary conditions for hydraulic and hydrologic modelling. In comparison with photogrammetric

methods that can only collect data and perform measurements from areas visible from several photographs, laser technologies can (not all lasers) penetrate vegetation and tree canopy and provide ground data to generate DEM, even under dense vegetation. Indeed, Lallias-Tacon et al. [60] explain the geomorphologic evolution of channels under vegetation cover in the Drome River in France, producing at the reach scale (< 1 km length for ~ 600 m width), a detrended DEM with vertical variability ranging from − 5 and 13 m. Arguably, it is thanks to the ability of the LiDAR to "see through" the vegetation that allows such detailed reconstruction of the topography, as other traditional surveys would take more than a lifetime to collect such dataset at a similar resolution and accuracy. The authors worked from airborne Riegl Laser Scanner LMSQ560 and VQ-480 providing point densities of respectively 6.4 and 3.5 points/m^2. This being written, LiDAR is not a silver bullet and depending on which environment and at which scale you are working (and depending on your budget), you may want to combine and choose different types of technologies.

In the likes of SfM-MVS, LiDAR technologies have been used from the river scale down to the scale of local features, such as a point-bar or a gravel surface. They offer the advantage of being able to acquire data through vegetation and also through water with green-lasers, breaking the boundaries between underwater and terrestrial features, in specific conditions. It ensures that fluvial geomorphological mapping has been greatly enhanced by LiDAR technology, for "static" morphological analysis [18, 70, 78] and also dor channel evolution. Evolution analysis is usually performed by comparing several LiDAR datasets with one another (e.g. [110]) or by comparing LiDAR data to other datasets even when matching different types of data can lead to relatively low-density point-cloud as shown by research on the Dijle River in Belgium, (1 point per 4 m^2) and the Ambleve river (5–8 points/ m^2, RMSE 0.17 m) where the authors used a dataset calibrated against 1040 GNSS GCPs [84].

Airborne LiDAR for floodplain and related structures, such as megafans, is a perfect example of the ability of the technology to generate topographic dataset over several kilometre scales, while still offering a vertical resolution of a few centimetres [80]. The authors imaged a surface of 400 km^2 on the Brenta River alluvial megafan in Italy, and they could construct a DEM with a z accuracy of < 5–10 cm, when compared with about 700 Differential Global Positioning System (DGPS) points. The DEM constructed from the point-cloud ([80]: Fig. 2) shows the present course of the Bacchiglione River flowing towards Padua City, as well as the paleolandforms, with the previous river course and the abandoned meanders cut in the megafan.

Point clouds have also been essential in floodplain geomorphology to improve the mapping of flood for flow simulations. For instance, Chen et al. [24] used airborne data of the Apalachicola River in the USA. The river flows towards the Gulf of Mexico in Florida in a very "flat" and low-gradient area. The floods were modelled with Hec-RAS, and therefore, accurate and HRT was essential. Furthermore, the meandering river is connected to numerous sloughs, making it a complex environment to model. It is the LiDAR data that provided the necessary accuracy to model and understand the water pathways through the interconnected abandoned meanders and other floodplain features. Similar results were also shown using LiDAR-derived DEMs of levee breach splays and crevasses due to the floods of the Vistula River [120].

If we zoom further at the reach scale, application of point-cloud technologies to flood mapping, simulation for hazards and disaster risk analysis and process investigations are then performed using TLS as well. For instance, using a Riegl LMS Z210 terrestrial laser scanner in 2003, Heritage and Hetherington [44] have scanned a channel reach of the River Wharfe at Deepdale in North Yorkshire (UK), merging scans from 21 scanner positions. The scans were merged with one another using a set of GCPs recorded with an EDM theodolite. The dataset comprises 17 million points covering the river channel and the banks area for an area of 150 m 15 m. The accuracy of the TLS data was 0.038 m mean and 0.1673 standard deviation for 157 points used for alignment. As the laser reflects back from different surfaces, pointed at different angles, those values change with the type of surface. For Bedrock, the authors recorded a mean error of − 0.0065 m for bedrock, 0.25 m for rock gaps, 0.0749 m on broad-leaved vegetation, 0.0686 m on grass, − 0.255 m on water and water edge. The standard deviation varies from 0.13 to 0.14 m for rock gaps and bedrock, broad-leaved vegetation and water, while grass and water edge present a standard deviation around 0.08 m. The scanner therefore offers different levels of accuracy depending on what is being scanned. But besides those small-scale errors that are not relevant to general geomorphology, laser technologies have proven themselves at unrivalled scale, and so as early as the mid-1990s: e.g. the acquisition of point clouds that are dense enough to extract information at the grain scale, defined as 4000–10,000 pts/m^2 [63].

Finally, if LiDAR can see through vegetation, both SfM-MVS and LiDAR record data about it, and it makes it an important set of tools to work on plant successions and types of vegetation divided by geomorphological units. Verrelst et al. [113] distinguished vegetation of sandy levees, pioneer vegetation on riverbanks, temporarily inundated pastures, shrubs, willow, poplar and dirt road along the Millingerwaard floodplain along the Waal River in the Netherlands, by combining LiDAR and a RTK-GPS with 2 cm vertical accuracy for instance.

4.2.2 Examples at Different Scales

Researchers and practitioners have therefore worked on floodplains and the fluvial corridor at different scales, and although it is impossible to cover every single aspect of this work, I am providing you in the following sections with examples of point-cloud datasets used in the fluvial corridor (a portion of the floodplain), at the river scale, the reach scale and finally at the river width scale (Fig. 4.1). Those divisions may not be the most pertinent ones in terms of geomorphology or hydrology, but they are a good division of the different types of methods used across those scales.

4.2.2.1 The River Scale

Difficulties	– Elongated thin strip with often limited vertical variation – Often associated with forested vegetation hiding the river and elements of the fluvial corridor – Seasonal water-level change modifies the geometry that can be captured with inundated zones being difficult to capture
Recommended tools (not exhaustive)	– Airborne LiDAR from traditional airplane – Fix-wing UAV LiDAR – In vegetation-deprived environments airborne platforms for photogrammetry work as well – Aerial photographs stitching (understanding the limitations)

At the river scale, and because of the high-proportion of vegetation-covered surfaces, LiDAR is certainly the most adapted method, but there are situations when you cannot shy away from the photogrammetric methods, and this is particularly true if you are trying to use historical photographs to reconstruct past-environment, as there is no possibility (yet) to go back in time and measure past landforms and vegetation. In such a case, working with photogrammetric methods on historical photographs is a good option, although historical photographs were not taken with the aim of creating HRT 3D data and numerous difficulties can arise.

From historical aerial photographs, SfM-MVS can be used to model river corridors for instance, and most especially for rivers where the vegetation does not fully cover the bed (Fig. 4.2). This method is of course not limited to historical data, and using 1483 photographs taken from a traditionally manned helicopter, Dietrich [33] captured the riverscape along river reaches of 380–4900 m long with a total of 1483 photographs for an announced overall cost of $2900, even after the helicopter and pilot-time rental. Providing that the channel water is clear enough and that the

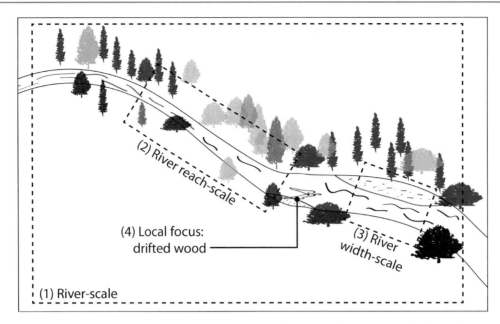

Fig. 4.1 Various focus areas where SfM-MVS has been employed, (1) at the river scale, (2) at a reach scale where a homogeneous portion of a river is examined and (3) at the river width scale where the length examined is equal to a few river width or even inferior to it, with point-bar and gravel examinations for instance

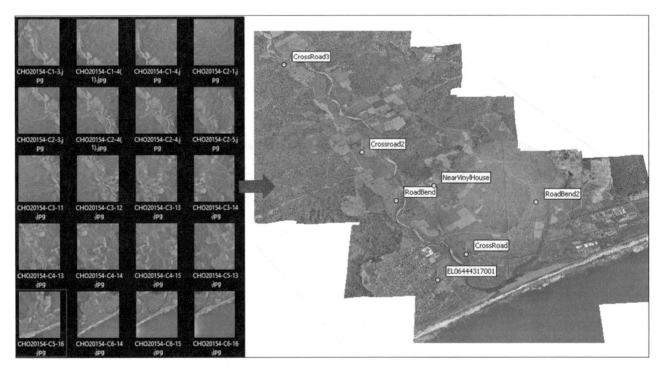

Fig. 4.2 From aerial photographs to 3D model using the Agisoft Metashape-Pro software using the aerial photographs from the Geospatial Authority of Japan (www.gsi.go.jp). The photographs were acquired in July 2015 at an altitude of 3230 m with a digital colour DMC2-230 camera of 92.041 mm focal. The blue flags are the ground control points (GCPs) with known coordinates (X, Y, Z). Please note that you will also need another set of points with known values to check for the error in the model

water surface illumination is right (i.e. without an overwhelming amount of Sun reflections at the water surface), it is even possible to use image spectral method to extract the stream depth from the SfM-MVS generated orthophotographs.

4.2.2.2 The Reach Scale

Difficulties	– Water-level change over time and water surface reflection – Vegetation hiding point-bars and details of the topography – Small-scale variations in the topography
Recommended tools (not exhaustive)	– Vehicle-mounted TLS – UAV SfM-MVS or LiDAR – Helicopter-based LiDAR

At the reach scale (i.e. a couple of hundreds to a few kilometres long), the needs for professional aircraft are not necessary anymore, and a low-cost UAV platform, either a quadcopter or a fixed-wing or a vehicle-mounted laser scanner, can be used to collect the 3D point-cloud dataset. Generally, airborne and terrestrial laser scanners will provide better results where vegetation is important, while UAV SfM will be more advantageous (and cheaper to buy and operate) for surfaces that are deprived of vegetation. At the same time, a dataset taken with this purpose in mind will provide a dataset that is sufficiently accurate for fluvial and floodplain geomorphology purposes (Table 4.3). And, authors have argued that SfM used in geomorphology is even having an important role in the field of river restoration [71].

As an example of a river reach scale example, I have drawn a point-cloud that was generated out of two sets of flights with a fixed-wing UAV and with a quadcopter UAV. The research project this dataset is built for, aimed at investigation the evolution of the geomorphology, especially looking at how this evolution was articulated with the bridge piles and the concrete road embankments, creating an area that is relatively fixed in the floodplain compared to the rest of the Quaternary sands and gravels. The first flight was calibrated spatially using a set of GCPs spread on the ground across the surveyed area (Fig. 4.3a), and they were surveyed with a GNSS Trimble R8 station and rover coupled in RTK over a radio link. The point-cloud then generated was used as the base point-cloud and the following point-cloud was calibrated from points of the first point-cloud, which are known for being perfectly static at the scene scale (buildings, road features, etc.). The surveys resulted in two point-clouds (Fig. 4.3b, c). In other words, there are environments, where very little elements are static and resurveying GCPs at each survey is necessary (and whenever possible, it is best to resurvey the same GCP location), but in cases where only a portion of your survey field is changing, it is a much more economic option, and easier option to measure GCPs of your first and last survey, and only calibrate the "mid-surveys" against a set of known points from the first survey. In case of a changing environment, you can use the change between a GCP_a at time t and the same one at time $t + \Delta$. This way, you have an idealized trajectory in time, and you can use this trajectory to create GCPs for any survey that is in between the two time steps. This is of course based on the assumption of linear change. This method starts to be "usable" at the reach scale, rather than the larger "river scale", because your point density starts to be high enough so that you can pick up points of your pointclouds to be at or almost at the GCP location you are interested in. If you have a pointcloud, with a density < 1 pt/m^2 for instance, linking the exact position of a given point at survey n, to survey $n + 1$ will require you introducing some form of interpolation and then further uncertainty. At the reach scale, I think—from experience— that your pointcloud is dense enough for the uncertainty trade-off to be advantageous enough to proceed with.

4.2.2.3 The Local Scale or River Width Scale

Difficulties	– Water surface reflection – Vegetation and wind in vegetation moving trees and elements measured
Recommended tools (not exhaustive)	– TLS – UAV-based SfM-MVS – UAV-based LiDAR – Vehicle-mounted LiDAR

At the local reach scale, that we have qualified in this non-standard division of space as being about one to a couple of width of the river metric, UAV-based photogrammetry using SfM-MVS allows thanks to the

Table 4.3 Camera, image capture specification and error metrics for ten ground control points used for the creation of one of the pointcloud of the Kowhai River at ~ 920 pts/m^2 presented in this subsection (the data was collected by the author and with the help of the technical team (Nicholas Key, Justin Harrison, and Paul Bealing) of the (then) Geography Department of the University of Canterbury—New Zealand)

Canon Powershot SX260	Focal length (mm)	Pixel size (mm)	Number of photographs	Mean flight altitude (m)	MAE (m)	RMSE$_Z$ (m)	RMSE$_{XY}$ (m)
	4.5	0.0015494	508	550	0.059	0.031	0.057

Fig. 4.3 Point-cloud generated by SfM-MVS from a fixed-wing platform of the Kowhai River in New Zealand, in October 2014 and August 2015.
(**a**) Representation in the dense point-cloud of one of the targets used as ground control point or for the measurement of error;
(**b**) view of the river reach downstream the highway bridge in October 2014 and in August 2015. The low-cost and easily reproducible survey is very opportune for rapid geomorphological change monitoring

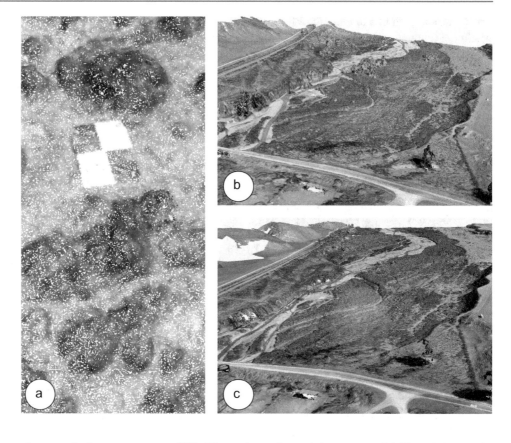

potentially low-speed of the quadcopter platforms to measure surface variation at a high resolution to eventually obtain the surface roughness of point-bars for instance, as it has been demonstrated for the Veneon river, in the French Alps [112]. The authors achieved with point density of 36–1052 pts. m^{-2} a roughness resolution between 0.3 and 1.3 cm with an average point distance of 0.1–6.0 cm (although those values are not inclusive of the potential error in the dataset). Despite of this potential difficulty, the authors reached a correspondence between the median roughness and the median grain size of $y = 0.42x + 7.62$ with a R^2 of 0.86, stating that despite of the potential errors in the dataset, the data still provides a proxy and a measure of actual grain roughness. Despite encouraging results, it has been shown for a 500 m long reach along the Thinhope Burn, a tributary of the South Tyne in Cumbria (UK) that several sources of errors generated by SfM-MVS could be found when compared with TLS data [99]. The authors have shown that the standard deviation of elevation error (SDE [m] tended to increase with the local surface elevation variation(LSEV [m] following the relation $SDE = 1.964 * LSEV^{0.49}$ with a R^2 of 0.77 [99]. This local error has been systematically analysed for different particle sizes, shapes, grain size structures and sorting, in order to better understand the error when attempting to extract a proxy of the grain size, especially because the recorded error has appeared to be site-specific

[87]. The authors showed the great variability of the results between studies as well as the existence of variable metrics: the local standard deviation and the median deviation from a plane (p. 145 in [87]). Investigating the role of gravel shapes on the relation of the gravel metrics with the SfM-MVS derived roughness, the following relations have been found: Table 4.4.

One of the important geomorphic factor that travels down the river corridors is the drifted wood, which can be quantified and measured using various methods, one of which is SfM-MVS. Furthermore, as drifted wood can generate expensive impacts on infrastructures, tracking its position and its eventual presence is essential and so at a low-cost. Offsetting both the travelling time of individuals through ground-mapping and the cost associated with LiDAR or traditional aerial photography, UAV-borne SfM-MVS appears to be a successful solution to those logistic issues. Research has shown that UAV-based SfM-MVS was effective at measuring volume of timber and wood jams, with a 13% discrepancy between traditional volume calculation from ground data and form SfM-MVS-built DSMs in the Blanco River off Chaiten Volcano in Chile [95].

Because, driftwood accumulation against road piles and other structures are an important element for engineers to assess, and because aerial vision is often obscured by the bridge deck for instance, ground-surveys can be needed, and

Table 4.4 Relations between particle shape (using D_{50}) and imbrication with the standard deviation of elevation (stdev) after the work of [87]

Imbricated prolate particles	$D_{50} = 6.51 * \text{stdev}—0.007$	$R^2 = 0.959$
Imbricated oblate particles	$D_{50} = 8.94 * \text{stdev}—0.005$	$R^2 = 0.966$
Prolate particles	$D_{50} = 9.63 * \text{stdev}$	$R^2 = 0.776$
Oblate particles	$D_{50} = 5.85 * \text{stdev}—0.001$	$R^2 = 0.988$

when a TLS is not a possible solution, ground-based SfM-MVS photogrammetry has proven to be able to deliver 3D volumes in combination with other methods from UAV and manned-aircraft [104].

4.2.3 Toolboxes for LiDAR Data Processing in Fluvial Geomorphology

From the availability of LiDAR data and high-resolution derivatives of different formats, the natural next-step has been the advent of dedicated toolboxes to automate some of the geomorphological processes. When I was a student, my main concern was to collect a sufficient amount of data during the time provided in the field, and as my fieldwork was occurring in Indonesia, time was of the essence before going back to the laboratory in France. Today, those questions have become more and more secondary, and the main concern is now centred on efficiently extracting meaningful information from datasets that are increasingly complex and voluminous, so that "doing it by hand" has become virtually impossible. One of those toolboxes, the TerEx toolbox, has been developed by Stout and Belmont [106], to extract semi-automatically fluvial terraces and other floodplain features from LiDAR. The TerEx toolbox was developed as an add-on to the ArcGIS® software, and the algorithm process is described by the authors as: "(a) select probable floodplain/terrace cells from a DEM, (b) eliminate selected areas that are not of interest, (c) generalize the shape of the selected areas and (d) [...] produce a shapefile consisting of the terrace and floodplain features" (p. 570 and cf. Fig. 1 page 571 for a flowchart of the algorithm). Testing their algorithm to find river terraces edges and other features over floodplains in USA and Australia, the predictor was successfully validated under different vegetation conditions, etc. Such standardized toolboxes are essential because the advent of low-cost UAV photogrammetry and laser systems has seen numerous developments in applied sciences and in various fields occupied by practitioners, such as post-disaster recovery and rebuilding projects (Fig. 4.4). In such a case, it is important to have standardized methods that can be directly applied and that can be compared between one location and another.

4.3 Selected Coastal Landforms and Adapted Point-Cloud Technologies

Continuing a downstream trajectory, following the sediment cascade and the pathway of water moving by gravity, in most cases, one will reach the coast. Coastal geomorphology is as complex as floodplain and fluvial geomorphology, and it also operates at a variety of intertwined scales. Along the coastline, the back-and-forth movement of waves make surveys even more complicated, as the waves will have gone several circles between the start and the end of a survey, would it be laser-based or photography-based.

And even if geomorphologists combine objects at different scales in their research, one can decompose this work in methodologies that are scale relevant. They often work on one or another landform before combining the approaches; and in this section, we are looking at some of the landforms that have been most studied using point-cloud technologies, to understand the different possibilities of Point-Cloud technology applied to coastal geomorphology. It will also be an astute way to identify areas that are still in need of work. Therefore, the reader should not look at the examples as an exhaustive review of coastal landforms.

4.3.1 Point-Cloud Technologies for Coastal Geomorphology

When it comes to Point-Cloud acquisition and processing, each environment has its set of challenges. Steep slope environments of the mountains present challenges to the acquisition and processing of point-cloud data due to the slopes' angles and often "closed landscapes" with limited visibility. One also needs to carry the instrument upslope, and it can be a challenge with a TLS of 40 kg for instance. For floodplains for instance, the challenges are different, landforms are usually smooth and spread over kilometres. Furthermore, vegetation and other land covers combine to further hamper geomorphological data acquisition. In the coastal environment, all those challenges are present together (steep slope cliffs, low variation spits and barriers); set of challenges to which one needs to add cyclic variations of the sea level, modifying what can be "seen" of the landforms.

Fig. 4.4 SfM-MVS derived
point-cloud of a valley restoration
project in Japan after a major
debris flow and landslide series of
events in 2017

Coastal landforms and processes are thus complex and display a huge variety of topographic realities (Fig. 4.5), and the elements shortly presented here are the results of a choice readers may not agree with but hopefully will make the best of. Within such a diversity of landforms and processes, finding the right classification is a complex task, especially because coastal scientists have already provided different options, based on dominant processes for instance ([73] p. 8). As the topic of the present book is point-cloud and geomorphology, I have decided to segment the coastal landforms following acquisition and processing needs. This way, the student of geomorphology will be able to compare the landform he/she is interested in studying and compare it with the set provided here, and then she/he will be able to decide what are the best tools for the study.

One of the earliest coastal topographic survey using point-cloud technology occurred on coastal sand dunes in the USA, at the Cape Hatteras in North Carolina [122], paper in which the author states that despite the emergence of LiDAR data and LiDAR for coastal mapping, there were still at that time very little research on coastal and dune morphology. This early research was conducted on an area spanning 100 m × 200 m, using LiDAR flown by the NOAA and NASA in 1996 and 1997.

Compared to UAV LiDAR and TLS, which we will develop as the two next section, conventional LiDAR has the advantage of being able to cover wide areas and for the coast very long strips. In Norway for instance, 80% of the coast (about 2500 without the islands and the Fjords, but estimated to be about 25,000 km included all the small islands, inlets and Fjords) has been mapped using conventional LiDAR, which would not be feasible in anybody's lifetime with TLS and or UAV LiDAR. The extracted DEM has been used to define potential inundation zones due to climate change and other storm surges, for which detailed data at the country scale is necessary to define natural hazards and disaster risks strategy as well as insurance policies [11].

The organization of the rest of this chapter follows an ocean-ward division, with the "hard-stuff" first (cliffs and shore platforms) and then the sediments that cover it (coastal sand dunes and beach) before getting into features like sandbars, spits and estuaries.

4.3.2 Cliffs and Point-Cloud Technologies

Difficulties	– Sub-vertical features difficult to map conventionally – Existence of overhang hard to map in 2.5D and 2D
Recommended tools (not exhaustive)	– TLS – Ground and tilted camera on UAV, flying at different elevations

Cliffs are steep coastal slopes cut into the bedrock or hardened sediments (over geological scales) with a cliff face slope that exceeds the 40° and which, often, is often sub-vertical. Cliffs are present along ∼ 50% of the world coasts [126]—in other words, it is a predominant geomorphologic features. The two main processes that generate cliffs are either differential erosion or tectonic processes (faulting, uplift...). Cliffs often present overhanging sections that are created by the basal erosion by the waves. Once the overhanging part falls at the foot of the cliff, the material is then being weathered and/or transported. Storms play an important role in this process as they have enough energy to quarry the base of the cliff, notably by transporting rock fragments and other debris, which increase the abrasive

Fig. 4.5 Examples of different coasts worldwide. **a** The basaltic coast of Comoros island created by volcanic effusive activity (*Photo* Gomez, 2006); **b** the rocky coast of the Stockholm archipelago in Sweden, where the present geomorphology is an inheritance of glacier abrasion (*Photo* Gomez, 2017); **c** The coast of Java Island in Indonesia, where the coral blocks mixed with blocks from the beachrock have been interpreted as the results of tsunami transport (*Photo* Gomez, 2007); **d** a sandy pocket beach in the Bay Peninsula area, South of Christchurch in New Zealand (*Photo* Gomez, 2013)

power. The modalities of cliff recessions and erosion however vary greatly depending on the material they are carved from and so are the mean cliff-recession speeds, varying from 1 mm per year in hard granite to 10 cm to a year in chalk cliffs and more than 10 m per year in volcanic ash [108]. The morphology of cliffs changes and varies depending on the geological formation they are made from and the modalities of interaction with the erosive agents. The erosion of cliffs is not a rectilinear process, as it is triggered and prepared by wave actions [83] for instance, with sudden acceleration in the erosion rates during storms [42], and eventually tsunamis. For the coastal cliffs of Santa Barbara (CA, USA), for instance, Young [125] has reported erosion rates of 0.11 ± 0.015 m/year between 1998 and 2010. At the storm scale, Alessio and Keller [1] have measured erosion rates of the same stretch of coastal cliffs of 0.5–3.5 mm/day during three storm events in early 2016. With climate change-induced sea-level rise [69], the erosion and retreat rates of cliffs have been of particular interest in recent years (e.g. [1, 27, 68]), and point-cloud technologies play a central role in this assessment.

The morphology also holds important information on the longer-term behaviour of coastal cliffs. Bird [6] writes that "cliff morphology is three dimensional, and includes cliff profiles shaped at right angles to the coastline and cliff outlines in plan" (p. 74). He further writes that "cliff profiles on resistant rocks are generally bold and steep, becoming gentler on weaker outcrops" (p. 74). If the cliff becomes out of reach from coastal erosion, it then can evolve in a coastal bluff, that is a landform with a thicker sediment mantle and more "diffuse" shapes. Cliffs are consequently excellent candidates for 3D point-cloud use in geomorphology. Moreover, cliffs are vertical windows into rock and sediment formations, and point-cloud technologies have been fuelling the development of virtual outcrops in the field of geology [42, 45, 93].

ALS, TLS and cliffs—Unsurprisingly, there are relatively less work on coastal cliffs with ALS than there are for sand dunes and other features, because those are often sub-vertical and other point-cloud acquisition methods may be preferred, even if large coastal extents still benefit from the ability of ALS to capture long portion of coastal cliffs. For instance, Lidar multiband has been used to survey coastal cliffs in Australia [5]. By contrast, terrestrial laser scanner (TLS) differs from ALS based on the place it takes the data from: it is not flying one or a set of imaginary parallel above the ground, but is "stationary" in comparison with the airborne survey—or semi-stationary as it can be mounted on a kart or on a vehicle. This results in two major differences between the two point-clouds generated: (1) the TLS extent is more local, and (2) the TLS can provide data on complex topographies that are hardly visible from ALS. In the coastal environment, this is particularly important for the monitoring of coastal cliffs and other steep slopes, as they do not appear well from ALS data (Fig. 4.6).

Consequently, contrasting with the ALS data and its limitations over steep slopes, TLS data has shown to be an effective way to measure, survey and follow the evolution of coastal cliffs [118]. Using 24 TLS surveys from February 2015 and March 2017 combined with a UAV-flight for SfM-MVS purposes [118] measured the rockfall patterns along a coastal limestone cliff of 0.6 km in the Marsden Bay of Northern England using a Riegl LMS-Z620 acquiring 11,000 points per second at a 3D spacing of 0.05 m at 100 m distance. The vertical surface of the cliff was treated as a horizontal surface. Their work combined with the geological strata data shows that the volume of rockfalls generated in the brecciated limestone is much higher than the underlying dolomitic limestone notably (Fig. 4 in [118]). They further evidenced that over time, complex patterns of erosion were operating at different rhythms depending on the lithology, and the season: 0.01 mm/day mean erosion rate in summer against 0.04 mm/day in winter, and a maximum of 0.09 mm/day in summer and in winter for instance. This work shows very clearly that (1) high-resolution topography is essential to improve our understanding of rock formation roles in erosion patterns and rates, and that (2) seasonal variations of erosion are significant and can be decomposed at unprecedented shorter time scale. It is this kind of technology that could lead to erosion forecast like we have weather forecast, by then reducing the disaster risks associated with cliff collapses.

SfM-MVS photogrammetry—SfM-MVS calculated from UAV-based photographs encounter the same issue as ALS for cliffs. NADIR images only offer a very limited view of the sub-vertical surfaces, and tilting the camera, like it has been done in forestry [35]. From the experiments of Jaud et al. [52], the seemingly best results in constructing dense point-clouds were for a camera angle at 20°. The camera is a Nikon D700 mounted on a DS6 DroneSys hexacopter. The flight with the camera orientation in NADIR yielded the worst results, while the tests at 30 and 40° off nadir saw a decrease in the number of points constructed. The authors reported 5818 tie points in nadir, 21,718 at 20°, 10,114 at 30° and 8560 at 40°. The quality of the dataset was assessed from a TLS Riegl VZ-400. I would argue that it is a combination of the flight height and the tilt angle that matters and facing a cliff varying the distance as well as the height in front of the cliff to take the photograph is essential to limit the deformation, but information on the flightpath was not included in the article.

Hand-held smartphone only offers little variation in the Z position of the camera, and for coastal cliffs for instance, the camera is either at the bottom of the cliff or further away on the beach, but the lack of change in the Z position relative to the cliff is the origin of important error, which creates deformation and error that increase with altitude away from the altitude of the camera. This appears well in the tests run by Jaud et al. [52] on a French pocket beach and cliff, where the photographs were taken from the beach and without GCPs near the top of the cliff to correct the error. Furthermore, the authors did show that with distance from the target, the error was also increasing. There is therefore a difference between the smartphone SfM model and the TLS data, and there are discrepancies between the different SfM model. The authors also demonstrated that depending on how the photographs are taken (linear series of photographs parallel to the cliff, against what the authors named "fan-shaped", i.e. following a semi-circular motion expanding outwards) a difference of 0.5 m of the position of the cliff, and important discrepancies in the cliff shape could be found (Fig. 8 in Jaud et al. [52]).

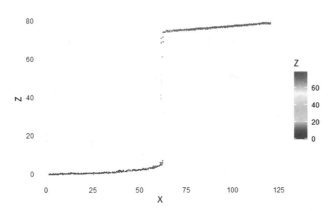

Fig. 4.6 Transect of the 2007 LiDAR from a cliff of the southern coast of the UK, illustrating the problems encountered when imaging sub-vertical surfaces from LiDAR data. The strip extracted is 5 m wide and extends between the two points X_0 [552413, 97286] and X_{end} [552471, 97384]

Comparisons and combinations of different techniques—
In an attempt to try defining what method works best to
define strike and dip from coastal cliffs, Cawood et al. [17]
compared the traditional compass, clinometer and their
digital version, to the results obtained from a RIEGL
VZ-2000 LiDAR, SfM from a camera pole with a Nikon
D3200, SfM from photographs taken from a DJI Phantom 3
UAV. From their experiments, terrestrial SfM provided the
best dataset to measure strike and dip, compared to
UAV-based photographs for SfM and laser data. They
attribute this unexpected result to the possibility to bring the
hand-held camera close to the target in a way that tiny details
can be captured. A promising approach to study coastal cliffs
both underwater and above is the combination of bathymetric LiDAR with multibeam sonar, as it has been operated
in Australia [54].

4.3.3 Coastal Dunes

Difficulties	– Large-scale features but needing high-resolution data – Constantly in movement, even in "seemingly" stable zones – Survey in windy areas transporting winds at the surface can lead to extensive error in the laser signal
Recommended tools (not exhaustive)	– ALS – UAV-borne SfM

Coastal dunes are asymmetric ridges or series of ridge
landforms generated by sand blowout from the beach, often
concentrated at the coast, where the sediment supply from a
sandy beach—sufficiently wide to allow the sand to dry and
then be transported by Aeolian processes. Coastal sand
dunes differ from the desert ones as they are not purely
Aeolian landforms. They are also influenced by the sea
waves during storms and by tsunamis [53]. Accurate topographic data obtained in a timely manner is essential because
coastal sand dunes are mobile landforms [36] and their
morphologies change under the combined actions of storms,
tsunamis, sea-level rise [32, 53] and vegetation cover [94],
with planting strategies often used to increase their heights
[50]. Sand dunes are also resilient structures, as they have
the ability to recover from erosional crises and return to a
dynamic equilibrium [26]. Coastal sand dunes are home to
rich and diverse ecosystems in equilibrium with the local
climate [34], and they also provide a natural barrier against
processes environmental hazards [75], except for large tsunamis for instance that can erode sand dunes to their roots as
it was the case along the coast of Lampuuk to the south of

Banda-Aceh City [53, 67]. A review on the coastal sand
dunes literature has shown that there is a worldwide trend of
dune stabilization, under the effects of several anthropogenic
factors, namely (a) land cover and land use change, (b) dune
stabilization projects; (c) decline in sediment supply; (d) and
change in the weather characteristics [36]. The study also
confirms the work of [48] that concluded that a global
"coastal greening" was occurring. This process is however at
odd with increased tourism in coastal areas that has led to
erosion and alteration of the sand dunes, then reducing their
ecological functions and biodiversity, for which LiDAR,
TLS and UAV-based technologies have proven crucial
[109]. Point-cloud technologies are therefore well indicated
to survey coastal dunes, their evolution.

ALS and TLS for Coastal Dunes
Woolard and Colby [122] stated that point-cloud technologies are essential to develop a comprehensive view of coastal
dune landforms and evolution processes. Even 2 m × 2 m
horizontally gridded topographic data is too coarse to
properly define the morphology of coastal sand dunes.
Coastal dunes being roughly topographic variations from a
horizontal plane, ALS has proven more effective than TLS in
acquiring data in an effective manner. Airborne LiDAR
surveys has been used to monitor dune evolution over time
[29], and it has also been used to investigate more "local"
processes, such as sand dunes blowout and depositional
lobes measurement using LiDAR differentiation between
2011 and 2014 [80].

In comparison with the steep slope cliffs, coastal dunes
are less propices to the use of TLS, and ALS arguably
provides the best ratio of area covered vs spatial resolution,
down to a certain resolution. For instance, using a
tripod-mounted Riegl VZ-2000 from December 2015 to
September 2017, a total of 22 surveys conducted at low tide
with an angular resolution fixed to 0.03°, with rescan at
0.002° resolution on the target reflectors were used to
characterized coastal sand dunes [28]. The advantage of the
fixed TLS position is its ability to virtually infinitely rescan
the same area to increase point density, or to increase the
angular resolution. This allowed the authors to work on
subunits of the coastal dunes, generating descriptors of the
dune crest position, elevation, backshore-dune slope, beach
width, beach volume and backshore-dune compartment.

For local features of sand dunes, and "local" surveys of
sand dunes, TLS remains preferable however [76]. This
being stated, if you have access to any of the two technologies, the power of point-cloud data is to be able to break
the small-scale = small features, large-scale = large features
rules, and there is certainly a lot to be learned by working
from a large data sample.

Combinations and Comparisons of SfM with ALS and TLS for Coastal Dunes

Sand dunes show sufficient vertical variation for SfM-MVS which can be used with proper ground control points pinged with a GNSS in RTK or PPK. It is less expensive, but the ephemeral nature of the coastal landforms will certainly need you to have a more expensive control tool to calibrate your point-cloud. One example of such work has been done from the Garopaba Dune field, where Gohmann et al. [39] have combined the use of SfM from UAV platform with ALS and TLS over a period of 2010–2019 to investigate the migration of coastal sand dunes. From their work, the migration rate of the sand dune being at a rate of about 5 m per year during nine years was extracted. The author further explained that while the TLS took three days of fieldwork and eight days of processing, UAV SfM data acquisition took under 3 h with a total station to map the control points. The time window you can allow for your data acquisition, combined with the necessary precision and accuracy you need are thus all elements you need to factor before deciding which method to use.

4.3.4 Beaches

Difficulties	– Long strips with limited topographic variation – One constantly moving limit (where to put it anyway) – For photography matching, the constant change in wetting pattern or wave rolling in and out makes SfM-MVS difficult;
Recommended tools (not exhaustive)	– ALS – TLS

Beaches are accumulation of sediments, mostly between sand and gravels sizes, that is spatially limited landward by the upper limit of the swash and seaward by the lowest tide level, although those limits are the object of debates and controversies, but it gives an idea of the spatial band between land and sea that the term beach defines. Even if I would rather leave this debate to experts, it is important to acknowledge that it will have an impact on how we measure and use point-cloud technologies on beaches. The material that forms them can have multiple origin, from river sediments to coastal cliff erosion and submarine sediment transport. Under an apparent simplicity, beaches are therefore very complex with sediment sizes that can change as river sediment-discharge changes and with density that can increase several folds from the calcareous shell to the quartz grains they can be made of.

Because of the low topographic variation in z over large horizontal scales, the error generated by photogrammetry is often too large to be employed on beaches [12]. DEMs of beach shorefaces are thus best generated from TLS and ALS [105], because accurate morphometric measurement is essential to understand the processes shaping it. Because of the specificities mentioned above, laser scanners are also sometimes mounted on poles to collect a series of 2D transects at high temporal resolution [2]. Using this technology, Andriolo et al. [3] could measure changes between − 0.2 and + 0.2 m within 24 h to a distance up to 50 m from the sensor (a SICK LMS500).

The importance of TLS and ALS measurement of beaches has been further brought to light due to recent concerns over the impacts of climate change-induced sea-level rise, increased wave energy and the resulting general erosion of the shorelines, as it can be seen in California for instance (https://www.climateassessment.ca.gov/). LiDAR technology provides the tools to follow the resulting changes both spatially and temporally. Using six consecutive LiDAR flights over the coasts of California between 1998 and 2016 (vertical uncertainty between 0.024 and 0.15 m), Smith and Barnard [101] have notably measured the impacts of El Nino's abnormal high wave-energy and erosion. On Chinese coasts, Ge et al. [38] used TLS to make a similar measurement of the impacts of storm surges. Away from the ocean and seas, beaches around smaller water bodies, such as glacial lakes also, provide essential information on past-climate changes. Yang and Teller [123] measured the shoreline succession of Lake Agassiz near the northern border of the USA. The succession of glacial lake beaches provides important information on the evolution of the paleolake for instance.

4.3.5 Spits and Coastal Barriers

Difficulties	– Long strips with limited topographic variation – One to two constantly moving limits (where to put it anyway) – For photography matching, the constant change in wetting pattern or wave rolling in and out makes SfM-MVS difficult
Recommended tools (not exhaustive)	– ALS – TLS

In terms of morphology, spits and coastal barriers do not differ much from the beach, as they are sedimentary accumulation above the high-tide level, but they differ because they are built in between a body of water inland and the sea, the inland body of water being of different nature. The spits are also usually generated by sediments drifted along the coast, so that the planform of their tip often resembles a young fern-leaf. Coastal barriers are generated by a broader range of processes and although they can be reduced to one single beach, they can be several hundred metres wide with coastal dune formations over it. These two landforms can in turn be differentiated from bars, which are submerged for part or all the tidal cycle. Bird [6] groups all the barrier systems together, and differentiate them based on their morphology (no free ends/one free end/two free ends) and for those with one free end, he divides them between spits and cuspate forelands (after Ollerhead 1993 in Bird [6]).

4.3.6 Submerged and Semi-submerged Landforms: Shore Platforms, Estuaries, Deltas, Coastal Marsh and Wetlands

Difficulties	– Several kilometres' horizontal scale for small vertical variation – Partly or fully underwater
Recommended tools (not exhaustive)	– Bathymetric Lidar – Multiband TLS (local scale) – Underwater camera for SfM-MVS (local scale)

Going to the limit of what LiDAR can image, there is at the coast a group of landforms that are challenging because they are not purely submarine features and they are not purely land features. This division is particularly difficult when it comes to point-cloud technologies, but LiDAR has proven to work well for shallow coastal bathymetry (Fig. 4.7). Among those, we can find shore platforms, estuaries, deltas and the coastal wetlands, which all present a number of challenges.

Shore platforms—As a rule of thumb, when the environment is depositional (subsidence, low-energy wave climate, etc.), sediment accumulation creates the coasts prized by the holiday makers, but uplift for instance will replace the sediment accumulation by a rocky sub-horizontal surface, often in between cliffs and the sea: the shore platform. Those surfaces are usually irregular with potholes generated by wave action combined with rock debris, which can be quarried from shore platform or the coastal cliff. The alternation of wetting and drying with the tides contributes to rock weakening and freeze–thaw also occurs under cold climates. The mechanical effects of waves and direct atmospheric processes at low tide are accompanied with chemical weathering varying with the nature of the rock.

Estuaries and deltas—Estuaries are coastal inlets shaped by erosion and deposition, balancing the effects from the river, the waves and the tides (estuaries have been divided between tide-dominated and wave-dominated) as well as the sediment supplies. Cameron and Pritchard [14] defined them as "semi-enclosed coastal water bodies having a connection with the open sea and within which seawater is measurably diluted with fresh water derived from land drainage. This same definition was also used by Nordstrom and Jackson [81]. Estuaries and deltas are not easier to define, this is why they have been the object of dozens of definitions [88], and even more so because they evolve through time: estuaries can be open at time and close at other, depending on river discharge and sediment availability (e.g. the Balcombe and Merricks Creeks in Victoria, Australia [55]).

Coastal Wetlands—Coastal wetlands are transition zones between the marine and the terrestrial environment, stretching as a "band" rather than being a linear limit. They are often a mix of emerged and submerged land, with a proportion that varies with the tides, season, etc. Despites the numerous difficulties, the bio-ecological interests of those zones has attracted a number of LiDAR-based research.

4.3.7 A Few Words on Managed and Degraded Landforms

One temptation here could have been to write a section on man-made landforms (Fig. 4.8), but from the perspective of pointcloud technologies—epistemologically speaking, there is no major difference between the two, and ontologically this simple (Devil vs. God) Judeo-Christian concept of our relation with nature is hardly relevant when it comes to coastal geomorphology. In other terms, Nordstrom writes in his book on developed coastlines that "natural landscapes are a myth that human agency is not an intrusion but part of the coastal environment and that human-altered landscapes can and should be modelled as a generic system […]. It is difficult to obtain the proper level of objectivity when static natural systems are considered the standard by which human-altered systems are evaluated. Terms used to describe human modifications to shorelines include "intervention", "disturbance", "alteration" and "bias". It would appear that the term "alteration" is the most appropriate of these terms if humans are considered intrinsic agents" ([82], p. 265). It is not the goal to discuss the Judeo-Christian ethics of the relation between man and nature, here, but Nordstrom formulates very clearly one of the main issues of restoration in geomorphology: the lack of knowledge of what is the original state, especially when the original state is a dynamic equilibrium that evolves from various anthropogenic and natural forcing. One can only rejoin Nordstrom on his conclusion that scientists are as likely to find a state of origins as they are to find the Garden of Eden from scientific investigations. This blanket conclusion is especially true for the coastline, because it was long seen as a low-interest area. This being stated, it should not then be an open door for all

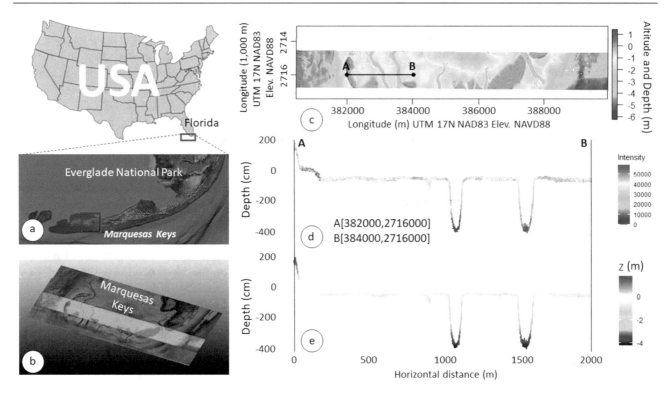

Fig. 4.7 LiDAR imaging of the Marquesas Keys (USA), with **a** the location to the South of the state of Florida, **b** the strip of LiDAR data used; **c** the transect through the topography and bathymetry data; **d**, **e** topographic transect with the return intensity varying with depth underwater and a coloured depth representation to show the linkage between the two

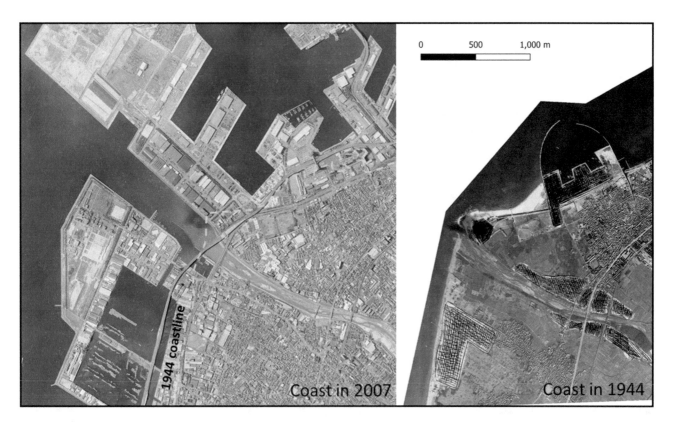

Fig. 4.8 The coast, South of Osaka-City in Japan showing the present-day coastal development and reclaimed land compared to the coast in 1944. In less than half of a century, the coastline has progressed seaward by several kilometres. The deep sediment layers in the bay are however a constant battle for engineers, in the image of the Osaka-Kansai airport, built on reclaimed land that is still sinking

kind of development and mismanagement. In acknowledging that humans are part of Nature and part of the environmental agents that inflect the trajectory of coastline morphologies, we also have a duty in interacting with those landforms in such a way that biodiversity and existing ecosystems are not brought to the brink of extinctions, and that development strategies do not become disasters for forthcoming generations (in the same way we pumped fossil fuel born greenhouse gases in the atmosphere).

Avoiding being trapped into an "image d'Epinal" of the coast (*meaning a naïve and rather-conservative stereotypical view of the coast in French*), it is important to therefore state that interaction between humans and the coast isn't neutral. For instance—and although it is still being used extensively along numerous coastlines—scientists and engineers know that replacing coastal materials (e.g. sands and gravels) with concrete is not a viable nor a sustainable management solution. Even in the nation that has arguably built most of its coasts as a fortress against the sea, climate change may undo a predicted quarter of 600 years of human engineering [47]. The fatality of a situation we have been drowning ourselves into is now contrasted with efforts towards bio-engineering and restoration, as it can be seen in the delta of the Mississippi River, under the impulse of the Obama administration. It is a battle that has been fought by coastal scientists—maybe not engineers—for decades already. I can still remember one of my colleagues, Dr. Deirdre Hart, strongly advocating the importance of natural buffers, such as sandy shores and sand and gravel beaches to absorb wave energy instead of concrete walls that just reflect the energy and at the same time generate erosion at the foot of them. This would typically be the case of one shooting oneself in the foot.

The approach taken in the present contribution on managed and constructed landforms is to state that there are degraded environments and perennial environments, the first one signifying that the environment is losing some of its ecological, environmental functions and the corresponding landform may just disappear over time as a result of human intervention, while the second is the opposite, and the idea of perennial signifies that the landform is in a dynamic equilibrium that will see an evolution, change or disappear into a different environment, but at an often slower pace than the one inflected by human interventions giving sufficient time to the environment and the ecological complex to evolve and mute. This recognizes that humans are part of the coastal ecosystem, but that they have ethical responsibilities towards the environment and the future. Pye and Allen [92] state that "Fundamental to a successful management strategy is an adequate understanding of the basic physical, chemical, biological and human properties and processes which affect coasts and estuaries, including their interactions and variability on different time and spatial scales" ([92] p. 1). Geomorphology and geomorphologists are therefore in a favourable position to address several of the coastal issues. If human intervention in nature often comes with a negative connotation in Western cultures, there are places where it is difficult to draw the line between what is natural and what is from anthropic management, and it is even more difficult to frame the human–nature interaction into negative or positive outcomes. Looking back at the coastal sand dunes defined above, the calibrated foredunes of the Aquitaine coasts in France are a typical example of collaboration between natural and human processes. They are the largest (> 30 m high) coastal dunes of Europe. "Their development is the result of important human intervention and accompaniment of natural processes since the middle of the nineteenth century" [80], and it is difficult to separate what comes from natural processes and from human activity.

The mobile nature of coastal sand dunes (and sand dunes in general) put them at odds with sedimentary urban developments that can have negative effects on their persistence [72].

4.4 Worked Example with R: A Meander on the Alabama River

If you are working at the meander level, one common task often involves the quantification of the meander size, meandering planform pattern, its evolution and how the planform patterns and sediment distribution come into play with the vegetation for instance. If you have created a point-cloud using SfM-MVS or using a LiDAR dataset, you are most likely to have a dataset in Cartesian coordinates. Working on a given meander or a set of meander, it might then be judicious to change from a Cartesian into a polar system, in such a way that the r of the polar coordinate corresponds to the radius of curvature (i.e. the distance from the outside of the bankfull channel to the centre of a circle. The centre of this circle of radius r is located at the junction between the bisecting segments that are perpendicular to the tangent lines of the curve starting points. As the riparian line is often difficult to define, it is possible to do the same work using the centreline of the bankfull channel. You can then use the angle θ of the radius of curvature to determine changes in the meander. Please note that if you choose to work on several meanders, it is important to choose the centre of the circle from a survey that occurred at the same date or period. As the meander will migrate, the position of this centre will change, and depending on the data you are interested in, it might be necessary to move the centre point over time or keep the centre point of a reference meander to compare it with future and past meanders (Eqs. 4.5 and 4.6).

$$r = \sqrt{(x^2 + y^2)} \qquad (4.5)$$

$$\theta = \tan^{-1}\left(\frac{y}{x}\right) \qquad (4.6)$$

If you are using the package LidR, which we are using through this book, you will need to first identify the location of the centre of your circle defined by the radius of curvature and then decide at what angle θ interval you want to sample your meander, and what is the length of the radius you want to sample (in the present case, we want to get transects that take the bankfull channel but also the surrounding floodplain).

\# For the present example, I downloaded data from a Meander of the Alabama River, recorded in 2017 by the U.S. Geological Survey and provided free of charge as a set of ".las files". Opening the four tiles that were downloaded in CloudCompare shows the different swaths of the LiDAR (Fig. 4.9), and once merged with the vegetation, you have your starting dataset (Fig. 4.10).

\# Once you have a dataset you are happy to work with (note that the previous operations can be done in R,

but on my system CloudCompare works faster with a large dataset than R, so that I prefer to prepare my data in CloudCompare before doing further processing in R. I think that it is just a matter of finding what kind of solution does work best for you, rather than an imperative).

\# In R, you should first load the libraries you will be using, and then we can open the dataset, and we are going to simplify the name to a standard las, for the las file

> library(lidR)
> las <- readLAS("USGS_LiDAR_Meander_Alabama Riv.las")

\# to make sure the data loaded properly, you can perform a simple plot. In the present example, I am plotting the intensity as a color gradient, and I chose to display both the X and Y scales (in meters) and the color gradient (Fig. 4.11). In the present case, we are working on a meander roughly 2500 m × 2000 m

> plot(las, color = "Intensity", bg = "white", axis = TRUE, legend = TRUE)

\# Using the function print(), for a las file, you can get the details of the data, in which case here we have the extents of the corner of the quadrangle (you can input those in your favourite Earth browser and find the location of the meander on the Alabama River), and we can see that we have slightly more than 24 million points, with a density which is a bit shy of 4 points/m^2

> print(las)

class:	LAS (v1.4 format 6)
memory:	2.4 Gb
extent:	423608.1, 426000, 3460701, 3463338 (xmin, xmax, ymin, ymax)
coord. ref.:	NAD83(2011)/UTM zone 16N + NAVD88 height—Geoid12B (metre) (with axis order normalized for visualization) (with axis order normalized for visualization)
area:	6.2 km^2
points:	24.34 million points
density:	3.93 points/m^2

Fig. 4.9 Visual representation of the LiDAR swath. Each colour corresponds to one flight and the corresponding LiDAR scan. You can note the areas where the LiDAR is overlapping, this will usually create areas where you have higher points densities, but at the same time, as you are close to the edge of your swath, this is where you may have more error to correct if you have to do the corrections yourself

\# If your point-cloud is not already classified when you use it (the USGS las file has already been classified when used), you can use the following function to create the ground (also part of the lidR library):

Fig. 4.10 Perspective view of the LiDAR data for a meander of the Alabama River in USA, collected by the USGS and provided as a laz file by the USGS free of charge

Fig. 4.11 Representation of the point-cloud using the intensity scale in R using the LidR package (please note that the point-cloud can be rotated on the screen and zoomed on as it is an interactive viewing environment)

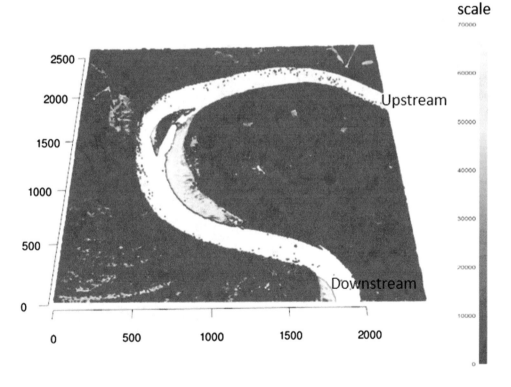

```
> las2 <- classify ground(las, algorithm = pmf(ws =
5, th = 3))
```

otherwise, you can use the preprocessed data and create a sub-dataset or work only from the points that are attributed to the ground. If you decide that you do not need the vegetation for your work, I would strongly suggest that you subset your data, so that there are less points in it. This will accelerate all the processing work. In the present case, the same area has now 7.3 million points once the vegetation is removed, instead of > 24 millions

```
> GroundOnly <- filter_ground(las)
> plot(GroundOnly, size = 3, bg = "white", axis =
TRUE, legend = TRUE)
```

using these commands, you can then get the Fig. 4.12 (please note that R will put the color scale on the right of the drawing, I just moved it in a horizontal position manually when doing the figure). Once you have this dataset, you are now ready to create transects along the radius of curvature.

For the present example, we are running one of those radius, from X_1 425145 Y_1 3462059 to X_2 423678 and Y_2 3462517, and to do so, we are using the method provided by Jean-Romain Roussel, Tristan R. H. Goodbody and Piotr Tompalski in the explanatory

book of their package (see the Cran website of the lidR package and the additional information and code on github)

```
> writeLAS(gnd,    "AlabamaGround.las",    index
= FALSE)
```

Finally, you may want to save your classified point-cloud of the ground, and to do so, you just need to use the following command, and it will save your point-cloud inside your working folder (please note that this command also requires you to load the lidR library):

```
> library(ggplot2)

> startpt <- c(425145, 3462059)
> endpt <- c(423678, 3462517)
> transect01 <- clip_transect(gnd, startpt, endpt,
width = 4, xz = TRUE)
> ggplot(transect01@data, aes(X,Z, color = Z)) +
    geom_point(size = 0.5) +
    theme_minimal() +
    scale_color_gradientn(colours = height.colors
    (50))
```

These few lines that were adapted from the presentation on Github of the authors of the lidR library create the graphic used as a base of Fig. 4.13

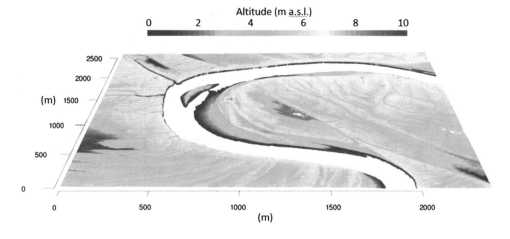

Fig. 4.12 Classified LiDAR point-cloud representing only the ground by altitude once the elevation has been removed

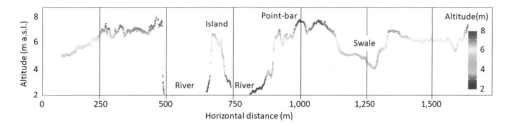

Fig. 4.13 Transect across the Alabama River. The choice of the angle of the curvature radius was chosen to coincide with the island visible on Fig. 4.12. This location, loosely termed island, is more certainly the emerging part of a point-bar or a sand-bar rather than a proper geomorphological island, but this way it is easy to understand

4.5 Worked Coastal Example with R

In this subsection, we will look at some simple code to help you get started with your project. The examples provided here have all been created for coastal dataset, and the R examples requires you virtually no prior knowledge of R, and after replicating the examples, you should be able to do this work on your own and modify the code easily for your own purposes. The task we are looking into are two essential processing in geomorphology: the extraction of a cross section, and the extraction of points to a grid and the change of spatial scale.

4.5.1 Extracting Cross Sections in R with the LidR Package

In this exercise, we are downloading data from the UK LiDAR repository, importing it in R, using the LidR library, and extracting the cross section, which can then be displayed using the ggplot library. This exercise follows a similar exercise proposed with the LidR package, and the reader should refer to the tutorials available with the LidR documentation to go further (Fig. 4.14).

```
library (raster)     # necessary for lidR
library (sp)         # necessary for lidR
library (lidR)       # import the lidR library to your
                       project
library (ggplot2)    # the graphic library for plotting
                       the transects
```

```
# you then want to make sure that you are in the right
working space, where your data is located, etc., so that
you are all nice and cozy. Please note that this is just a
setup on my computer, yours will look different. In
R-studio, you can also choose your workspace in the
files/more/option.
```

```
setwd    ("C:/Users/kaiki/Desktop/DESKTOPFILES/
LIDAR processing with lidR")
```

```
# read the pointcloud, which is a file I just named
"points" in this case, and that is accessed in R by using
the "las" variable.
```

```
las <- readLAS ("points.las")
```

Fig. 4.14 A 3D interactive plot of the pointcloud of the sea waves from the 2007 LiDAR data near New Brighton in the UK (*source of data* Department for Environment Food & Rural Affairs of the UK, https://environment.data.gov.uk/)

You can generate a 3D view of the pointcloud using the plot command in R with simply the plot function if you point to your 3d pointcloud.

```
plot (las)
```

you can also create transects between two points with the ggplot2 library. You need to first define two points on the *x*, *y* plane to draw a blade that will cut through your topographic data: these are the tr1_start and tr1_end variable. Then we save the data in between the two points in the "transect" variable. This variable, needs to include points at a given distance from the slice we are creating. In this case, as my pointcloud is in meters, and I chose width = 2, I am creating a 2 m wide slice.

```
tr1_start <- c(551464, 96997)
tr1_end <- c(551856, 97346)
transect <- clip_transect (las, tr1_start, tr1_end, width = 2, xz = TRUE)
ggplot (transect@data, aes (X, Z, color = Z)) +
geom_point (size = 0.5) +
coord_equal() +
theme_minimal() +
scale_color_gradientn (colours = height.colors(10))
```

this create a transect with all the points in a 2 m radius from the file las, from point 1 to point 2, and the color gradient is calculated according to the elevation. In the LAS file you have the intensity…etc. and even the classification in some cases, you can use those to modify the color scale. You will then get a transect similar to the one I presented in this chapter from the Florida dataset.

If you have difficulties finding the coordinates of your dataset, you can use the summary() function, which in the case of the coastal waves gave the following result:

```
summary (las)
```

class:	LAS (v1.2 format 2)
memory:	9.3 Mb
extent:	551463.9, 551856.6, 96996.27, 97346.47 (xmin, xmax, ymin, ymax)
coord. ref.:	NA
area:	137276.1 units^2
points:	106.2 thousand points

density:	0.77 points/units^2
File signature:	LASFÿ
File source ID:	255

Global encoding:

– GPS Time Type: GPS Week Time
– Synthetic Return Numbers: no
– Well Know Text: CRS is GeoTIFF
– Aggregate Model: false

Project ID—GUID:	00000000-0000-0000-0000-000000000000
Version:	1.2
System identifier:	PDAL
Generating software:	PDAL 2.1.0 (828987)
File creation d/y:	7/2021
header size:	227
Offset to point data:	473
Num. var. length record:	1
Point data format:	2
Point data record length:	34
Num. of point records:	106243
Num. of points by return:	106243 0 0 0 0
Scale factor X Y Z:	1e-06 1e-06 1e-06
Offset X Y Z:	551464 96996.3 −2.5
min X Y Z:	551463.9 96996.27 −2.5
max X Y Z:	551856.6 97346.47 −0.84
Variable length records:	

    Variable length record 1 of 1
        Description: Extra Bytes Record
        Extra Bytes Description:
            Original cloud index:

4.5.2   Extracting Coastal Data on a Grid in R with the LidR Package

In this sub-section, we are using once more data from the UK LiDAR database, and we are transforming the original point-cloud into a gridded dataset. In the present case, we are using the minimum value in Z among all the points in one cell of the grid. Please note, that in each cell, you can virtually do any form of calculation you want from the LiDAR point-cloud (density of points, maximum values, median, mean, standard deviation … etc.).

Search for the lowest elevation for each grid of 5 m, then generate one own color scale before plotting it. Looking at the average density of point per square meter (0.77) from the summary() function, trying to use a square with a footprint that is too small.

```
minlev5m <- grid_metrics(las, ~ min(Z), 5)

myscale <-colorRampPalette(c("black", "blue","cyan",
"pink","white", "red"), bias = 1.5)
plot(minlev5m, col = greyscale(100)) #
result in Fig. 4.15
```

Conclusion

Both floodplain and coastal geomorphology are fascinating, due to their diversity of landforms, slopes and scales that interact with one another. This diversity gives rise to complex geomorphology, which can, depending on the position of the floodplain and coast. At the same time, both environments are particularly challenging to topographic mapping due to the large spatial extent and often limited vertical variation. Furthermore, the important topographic gradient (a steep river bank or a cliff) both offer another type of challenge to point-cloud acquisition. As they are sub-vertical, they offer little surface to sample from aircraft photographs and laser scanners.

From this chapter, you should now have an idea of what kind of point-cloud technology you can use to collect data from one type of landform or another and understand the difficulties related to those. Furthermore, you should also be able to extract transects and grid data from a LiDAR (ALS or TLS) dataset in R using the dedicated LidR library.

Review Questions

– What are the difficulties encountered when using ALS, TLS and SfM on coastal and floodplain landforms? Give some examples and explain them from a geomorphologic standpoint.
– What is the most appropriate tool to record coastal cliffs retreat? Explain why? Find studies that have done so in the UK, and after downloading the LiDAR data of the studied areas, compare your data to the data used in the research you read.
– Why are 2D scanners on poles needed at the coast? What are the difficulties that the fixed scanners attempt to overcome?
– What is the limitation of SfM-MVS for floodplain mapping? In what kind of environment would it work best?

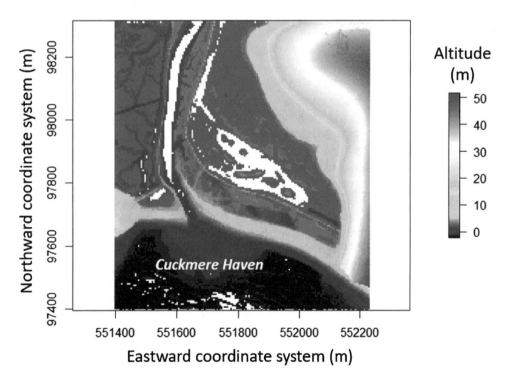

Fig. 4.15 Gridded Lidar data of 2007 of the Cuckmere Haven river mouth on the southern coast of the UK using the LidR package in R and R-studio

– Why are HRT pointclouds essential for floodplain topographic mapping?
– Explain how you could image engineered structures at the coast and recognize them from point-cloud data. What is the advantage of LiDAR over SfM-MVS?

Potential Problems for You to Reflect On

Imagine that you are working for a consulting firm in your home country or anywhere else, and you get contacted by a local authority from a coastal village. They are stating that they seem to have a problem of sedimentation at the coast due to what they think is the cause (dredging in a connected river and mismanagement of the watershed). They would like to quantify the problems they think they are observing.

Choose any location of your choice with coastal sediments, and download the LiDAR data from your favourite provider. Investigate at this location what is the sediment dynamic, and compare your result with weather data, with coastal current data if you have them, and with satellite imagery or aerial photographs showing the evolution of the watersheds, the strip of coast is connected to.

Furthermore, your manager tells you that your client wants access to the data, but they only accept GIS data in a grid format. Reflect on what is the right scale to produce the data. Are there differences in what you can show with different scale data. Explain the differences and make the difference between data artefacts and real changes at the coast (please note that depending on the coast you are choosing, you may see almost no change between two dates. This is a result as well, and it is fine as well).

As another example, think this time that you are the mayor of a small community along the Sacramento River in California, and you are briefing with your own in-house team of expert a plan to measure the effect of the 100 years flood on the floodplain of your community. Depending on where you are in the world, you may want to adapt this scenario to a close by area.

Using LiDAR data, extract cross section along a portion of the floodplain to create an input file for Hec-RAS (free) or any of your favourite flow routing software, and modify the sampling resolution of your dataset to 1 m to 10–25 m. Reflect on the importance of having a good topographic/bathymetric dataset in flow and flood modelling. Finally, does it make sense to invest huge amounts of money in a flow solver that works at a very fine scale when the bathymetric and topographic data is still at a coarse scale?

References and Suggested Readings

1. Alessio P, Keller EA (2020) Short-term patterns and processes of coastal cliff erosion in Santa-Barbara, California. Geomorphology 353(106994):1–12

2. Almeida LP, Masselink G, Russell PE, Davidson MA (2015) Observations of gravel beach dynamics during high energy wave conditions using a laser scanner. Geomorphology 228:15–27

3. Andriolo U, Almeida LP, Almar R (2018) Coupling terrestrial LiDar and video imagery to perform 3D intertidal beach topography. Coast Eng 140:232–239

4. Balaguer-Puig M, Marques-Mateu A, Lerma JL, Ibanez-Ascensio S (2017) Estimation of small-scale soil erosion in laboratory experiments with Structure from motion photogrammetry. Geomorphology 295:285–296

5. Bezore R, Kennedy DM, Ierodiaconou D (2019) The evolution of sea cliffs over multiple eustatic cycles in high energy, temperature envieonrments. Cont Shelf Res 189(103985):1–9

6. Bird E (2008) Coastal geomorphology—an introduction, 2nd edn. Wiley, Hoboken, 411p

7. Bistacchi A, Balsama F, Storti F, Mozafari M, Swennen R, Solum J, Tueckmantel C, Taberner C (2015) Geosphere 11:2031–2048

8. Brasington J, Langham J, Rumsby B (2003) Methodological sensitivity of morphometric estimates of coarse fluvial sediment transport. Geomorphology 53:299–316

9. Brasington J, Rumsby BT, McVey RA (2000) Monitoring and modelling morphological change in a braided gravel-bed river using high reolution GPS-based survey. Earth Proces Landform 25:973–990

10. Brasington J, Smart RMA (2003) Close range digital photogrammetric analysis of experimental drainage basin evolution. Earth Surf Proc Land 28:231–247

11. Breili K, Simpson MJR, Klokkervold E, Roaldsdotter RO (2020) High-accuracy coastal flood mapping for Norway using lidar data. Nat Hazards Earth Syst Sci 20:673–694

12. Brignone M, Schiaffino CF, Isla FI, Ferrari M (2012) A system for beach video-monitoring: beachkeeper plus. Comput Geosci 49:53–56

13. Brunier G, Michaud E, Fleury J, Anthony EJ, Morvan S, Gardel A (2020) Assessing the relationship between macro-faunal burrowing activity and mudflat geomorphology from UAV-based Structure-from-Motion photogrammetry. Remote Sens Envion 241(111717):1–17

14. Cameron WM, Pritchard DW (1963) Estuaries In: Hill MN (ed) The sea. Wiley, Hoboken, pp 306–324

15. Carr B (2019) Stromboli Volcano (Vents & Sciara del Fuoco), Italy. Distributed by OpenTopography, 10 Sept 2018. https://doi.org/10.5069/G9VX0DNG. Accessed 31 Dec 2020

16. Carrivick JJ, Smith WM, Quincey JD (2016) Structure from motion in the geosciences. Wiley Blackwell Publisher, Hoboken, 197p

17. Cawood AJ, Bond CE, Howell JA, Butler RWH, Totake Y (2017) LiDAR, UAV or compass-clinometer? Accuracy, coverage and the effects on structural models. J Struct Geol 98:67–82

18. Challis K (2006) Airborne laser altimetry in alleviated landscapes. Archaeol Prospect 13:103–127

19. Challis K, Howard AJ (2006) A review of trends within archaeological remote sensing in alluvial environments. Archaeol Prospect 13:231–240

20. Chandler JH (1999) Effective application of automated digital photogrammetry for geomorphological research. Earth Surface Proces Landform 24:51–63

21. Chandler JH, Ashmore P, Paola C, Gooch M, Varkaris F (2002) Monitoring river-channel change using terrestrial oblique digital imagery and automated digital photogrammetry. Ann Assoc Am Geogr 92:631–644

22. Chandler JH, Fryer JG, Jack A (2005) Metric capabilities of low cost digital cameras for close-range surface measurement. Photogram Record 20:12–26

23. Charlton R (2008) Fundamentals of fluvial geomorphology. Routledge, Milton Park, 234p

24. Chen YJ, Mossa J, Singh KK (2020) Floodplain response to varied flows in a large coastal plain river. Geomorphol 354. https://doi.org/10.1016/j.geomorph.2020.107035

25. Church M (2013) Refocusing geomorphology: field work in four acts. Geomorphology 200:184–192

26. Claudino-Sales V, Wang P, Horwitz M (2008) Factors controlling the survival of coastal dunes during multiple hurricane impacts in 2004 and 2005: Santa Rosa Barrier Island, Florida. Geomorphology 95:295–315

27. Collins BD, Kayen R, Sitar N (2007) Process-based empirical prediction of landslides in weakly lithified coastal cliffs, San Francisco, California, USA. In: Landslides and climate change: proceedings of the international conference on landslides and climate, Isle of Wight, UK. Taylor & Francis, pp 175–184

28. Conery I, Brodie K, Spore N, Walsh J (2020) Terrestrial LiDAR monitoring of coastal foredune evolution in managed and unmanaged systems. Earth Surf Proc Land 45:877–892

29. Darke I, Walker I, Hesp P (2016) Beach-dune sediment budgets and dune morphodynamics following coastal dune restoration, Wickaninnish Dunes, Canada. Earth Surf Proc Land 41:1370–1385

30. Davidson-Arnott RG, Law MN (1996) Measurement and prediction of long-term sediment supply to coastal foredunes. J Coastal Res 1:654–663

31. Davies T, Campbell B, Hall B, Gomez C (2013) Recent behaviour and sustainable future management of the Waiho River, Westland, New Zealand. J Hydrol NZ 52:41–56

32. DeWinter RC, Ruessink BG (2017) Sensitivity analysis of climate change impacts on dune erosion: case study for the Dutch Holland coast. Clim Change 141:658–701

33. Dietrich JT (2016) Riverscape mapping with helicopter-based structure-from-motion photogrammetry. Geomorphology 252: 144–157

34. Everard M, Jones L, Watts B (2010) Have we neglected the societal importance of Sand dunes? An ecosystem services perspective. Aquat Conserv Marine Freshwater Ecosyst 20:467–487

35. Fritz A, Kattenborn T, Koch B (2013) UAV-based photogrammetric point clouds—tree stem mapping in open stands in comparison to terrestrial laser scanner point clouds. ISPRS Int Arch Photogram Rem Sens Spat Inform Sci XL-1/W2:141–146

36. Gao J, Kennedy DM, Konlechner TM (2020) Coastal dune mobility over the past century: a global review. Prog Phys Geogr 44:814–836

37. Garde RJ (1970) Initiation of motion on a hydrodynamically rough surface; critical velocity approach. J Irrigat Power CBIP (India) 27:271–282

38. Ge Z, Dai Z, Pang W, Liz S, Wei W, Mei X, Huang H, Gu J (2017) LiDAR-based detection of past-typhoon recovery of a meso-macro-tidal beach in the Beidu Gulf, China. Mar Geol 391:127–143

39. Gohmann CH, Garcia GPB, Allonso AA, Alburquerque RW (2020) Dune migration and volume change from airborne LiDAR, terrestrial LiDAR and structure from motion-multi view stereo. Comput Geosci 143(104569):1–13

40. Gomez C, Hart DE (2013) Disaster gold rushes, sophisms and academic neo-colonialism: comments on 'earthquake disasters and resilience in the global North.' Geogr J 179:272–277

41. Gomez C, Kataoka K, Saputra A, Wassmer P, Urabe A, Morgenroth J, Kato A (2017) Photogrammetry-based texture analysis of a volcaniclastic outcrop-peel: low-cost alternative to TLS and automation potentialities using Haar wavelet and spatial-analysis algorithms. Forum Geografi 31. https://doi.org/10.23917/forgeo.v31i1.3977

42. Hapke C, Richmond B (2002) The impact of climatic and seismic events on the shortterm evolution of seacliffs based on 3-D mapping: northern Monterey Bay, California. Mar Geol 187:259–278

43. Henderson FM (1966) Open channel flow. MacMillan, New York

44. Heritage G, Hetherington D (2007) Towards a protocol for laser scanning in fluvial geomorphology. Earth Surf Proc Land 32:66–74

45. Hodgetts D, Drinkwater NJ, Hodgson J, Kavanagh J, Flint SS, Keogh KJ, Howell JA (2004) Three-dimensional geological models from outcrop data using digital data collection techniques: an example from the Tanqua Karoo depocentre, South Africa. Geol Soc Spec Pub 239:57–75

46. Houston J (2008) The economic value of beaches—a 2008 update. Shore and Beach 76:22–26

47. Inman M (2010) Working with water. Nat Clim Change 1:39–41

48. Jackson DW, Costas S, Gonzalez-Villanueva R et al (2019) A global 'greening' of coastal dunes: an integrated consequence of climate change? Global Planet Change 182:103026

49. Jackson NL, Nordstrom KF (2011) Aeolian sediment transport and landforms in managed coastal systems: a review. Aeol Res 3:181–196

50. James MR, Chandler JH, Eltner A, Fraser C, Miller PE, Mills JP, Noble T, Robson S, Lane SN (2019) Guidelines on the use of structure-from-motion photogrammetry in geomorphic research. Earth Surf Proc Land 44:2081–2084

51. Jaud M, Kervot M, Delacourt C, Bertin S (2019) Potential of smartphone SfM photogrammetry to measure coastal morphodynamics. Remote Sens 11(2242):1–17

52. Jaud M, Letortu P, Thery C, Grandjean P, Costa S, Macquaire O, Davidson R, Le Dantec N (2019) UAV survey of a coastal cliff face—selection of the best imaging angle. Measurement 139:10–20

53. Kain C, Gomez C, Wassmer P, Lavigne F, Hart D (2014) Truncated dunes as evidence of the 2004 tsunami in North Sumatra and environmental recovery post-tsunami. NZ Geogr 70:165–178

54. Kennedy DM, Ierodiaconou D, Schimel A (2014) Granitic coastal geomorphology: applying integrated terrestrial and bathymetric LiDAR with multibeam sonar to examine coastal landscape evolution. Earth Sruface Proces Landfrosm 39:1663–1674

55. Kennedy DM, McSweeney S, Mariana M, Zavadil E (2020) The geomorphology and evolution of intermittently open and close estuaries in large embayments in Victoria, Australia. Geomorphology 350(106892):1–14

56. Kondolf GM, Angermeir PI, Cummins K, Dunne T, Healey M, Kimmerer W, Moyle PB, Murphy D, Patten D, Railsback S, Reed DJ, Spies R, Twiss R (2008) Projecting cumulative benefits of multiple river restoration projects: an example from the Sacramento-San Joaquin river system in California. Environ Manage 42:933–945

57. Kondolf M, Piegay H (2003) Tools in fluvial geomorphology, 1st edn. Wiley, Hoboken, 688p

58. Kondolf M, Piegay H (2016) Tools in fluvial geomorphology, 2nd edn. Wiley, Hoboken, 548p

59. Labuz TA (2016) A review of field methods to survey coastal dunes—experience based on research from South Baltic Coast. J Coast Conserv 20(2):175–190

60. Lallias-Tacon S, Liebault F, Piegay H (2017) Use of airborne LiDAR and historical aerial photos for characterising the history of braided river floodplain morphology and vegetation responses. CATENA 149:742–759

61. Lane SN (2000) The measurement of river channel morphology using digital photogrammetry. Photogram Rec 16:937–961
62. Lane SN, Chandler J-H, Porfiri K (2001) Monitoring river channel and flume surfaces with digital photogrammetry. J Hydraul Eng 127:871–877
63. Lane SN, Chandler JH, Richards KS (1994) Developments in monitoring and terrain modelling small-scale river-bed topography. Earth Surf Proc Land 19:349–368
64. Lane SN, Westaway RM, Hicks DM (2003) Estimation of erosion and deposition volumes in a large, gravel-bed, braided river using synoptic remote sensing. Earth Surf Proc Land 28:249–271
65. Lane SN, Chandler JH, Richards KS (eds) (1998) Landform monitoring, modelling and analysis. Wiley, Hoboken, 454p
66. Lane SN, Hicks DM, Westaway RM (1999) Monitoring riverbed topography by digital photogrammetry with particular reference to braided channels. NIWA technical report, New Zealand
67. Lavigne F, Paris R, Grancher D, Wassmer P, Brunstein D, Vaultier F, Leone F, Flohic F, De Coster B, Gunawan T, Gomez C, Setiawan A, Cahyadi R, Fachrizal (2009) Reconstruction of Tsunami Inland propagation on December 26, 2004 in Banda Aceh, Indonesia, through field investigations. Pure Appl Geophys 166:259–281
68. Letortu P, Costa S, Maquaire O, Davidson R (2019) Marine and subaerial controls of coastal chalk cliff erosion in Normandy (France) based on a 7-year laser scanner monitoring. Geomorphology 335:76–91
69. Limber PW, Barnard PL, Vitousek S, Erikson LH (2018) A model ensemble for projecting multidecadal coastal cliff retreat during the 21st century. J Geophys Res Earth Surf 123–7:1566–1589
70. Lohani B, Mason DC (2001) Application of airborne scanning laser altimetry to the study of tidal channel geomorphology. ISPRS J Photogramm Remote Sens 56:100–120
71. Marteau B, Vericat D, Gibbins C, Batalla RJ, Green DR (2017) Application of structure-from-motion photogrammetry to river restoration. Earth Surf Proc Land 42:503–515
72. Martinez ML, Landgrave R, Silva R, Hesp P (2019) Shoreline dynamics and coastal dune stabilization in response to changes in infrastructure and climate. J Coastal Res 92:6–12
73. Masselink G, Hughes M, Knight J (2014) Introduction to coastal processes and geomorphology. Routledge, 432p
74. Maximiliano-Cordova C, Salgado K, Martinez ML, Mendoza E, Silva R, Guevara R, Feagin RA (2019) Does the functional richness of plants reduce wave erosion on embryo coastal dunes. Estuaries Coasts—Spec Iss Integr Ecosyst Coast Eng Pract 42:1730–1741
75. Montreuil A, Bullard J, Chandler J, Millet J (2013) Decadal and seasonal development of embryo dunes on an accreting macrotidal beach: North Lincolnshire, UK. Earth Surf Proc Land 38:1851–1861
76. Morgan JA, Brogan DJ, Nelson PA (2017) Application of structure-from-motion photogrammetry in laboratory flumes. Geomorphology 276:125–143
77. Neill CR (1968) Note on initial movement of coarse uniform material. IHR-IAHR 6:173–176
78. Nelson PA, Smith JA, Miller AJ (2006) Evolution of channel morphology and hydrologic response in an urbanizing drainage basin. Earth Surf Proc Land 31:1063–1079
79. Nicolae Lerma A, Bossard V (2020) Geomorphologic characteristics and evolution of managed dunes on the South West Coast of France. Geomorphology 367:107312
80. Ninfo A, Mozzi P, Abba T (2016) Integration of LiDAR and cropmark remote sensing for the study of fluvial and anthropogenic landforms in the Brenta-Bacchiglione alluvial plain (NE Italy). Geomorphology 260:64–78
81. Nordstrom KF, Jackson NL (2012) Physical processes and landforms on beaches in short fetch environments in estuaries, small lakes and reservoirs: a review. Earth Sci Rev 111:232–247
82. Nordstrom, KF (2000) Beaches and dunes of developed coasts. Cambridge University Press, p 338
83. Norris RM (1968) Sea cliff retreat near Santa Barbara, California. California Division of Mines and Geology. Miner Inform Serv 21–6:87–91
84. Notebaert B, Verstraeten G, Govers G, Poesen J (2008) Qualitative and quantitative applications of LiDAR imagery in fluvial geomorphology. Earth Process Landform 34:217–231
85. Pagan JI, Banon L, Lopez I, Banon C, Aragones L (2019) Monitoring the dune-beach system of Guardamar del Segura (Spain) using UAV, SfM and GIST techniques. Sci Total Environ 687:1034–1045
86. Pavlis T, Mason K (2017) The new world of 3D geologic mapping. GSA Today 27:4–10
87. Pearson E, Smith MW, Klaar MJ, Brown LE (2017) Can high resolution 3D topographic surveys provide reliable grain size estimates in gravel bed rivers? Geomorphology 293:143–155
88. Perillo GME (1996) Geomorphology and sedimentology of estuaries. Dev Sedimentol 53:471
89. Persendt FC, Gomez C (2016) Assessment of drainage network extractions in a low-relief area of the Cuvelai Basin (Namibia) from multiple sources: LiDAR, topographic maps, and digital aerial orthophotographs. Geomorphology 260:32–50
90. Piegay H, Kondolf MG, Minear JT, Vaudor L (2015) Trends in publications in fluvial geomorphology overy two decades: a truly new era in the discipline owing to recent technological revolution? Geomorphology 248:489–500
91. Purdie H, Bealing P, Gomez C, Anderson B, Marsh OJ (2020) Morphological changes to the terminus of a maritime glacier during advance and retreat phases: fox Glacier/Te Moeka o Tuawe, New Zealand. Geogr Ann Ser B. https://doi.org/10.1080/04353676.2020.1840179
92. Pye K, Allen JRL (2000) Past, present and future interactions, management challenges and research needs in coastal and estuarine environments. In: Pye K, Allen JRL (eds) Coastal and estuarine environments: sedimentology, geomorphology and geoarcheology. Geol Soc Lond Special Publications, vol 175, pp 1–4
93. Rarity F, Van Lanen XMT, Hodgetts D, Gawthorpe RL, Wilson P, Fabuel-Perez I, Redfern J (2014) LiDAR-based digital outcrops for sedimentological analysis: workflow and techniques. Geological Society, London, vol 387, pp 153–183
94. Ruggiero P, Hacker S, Seabloom E, Zarnetske P (2018) The role of vegetation in determining dune morphology, exposure to sea-level rise, and storm-induced coastal hazards: a US Pacific Northwest perspective. In: Moore LJ, Murray AB (eds) Barrier dynamics and response to changing climate. Springer, Berlin, pp 337–361
95. Sanhueza D, Picco L, Ruiz-Villanueva V, Iroume A, Ulloa H, Barrientos G (2019) Quantification of fluvial wood using UAVs and structure from motion. Geomorphology 345:106837
96. Saputra A, Rahardianto T, Gomez C (2017) The application of structure from motion (SfM) to identify the geological structure and outcrop studies. In: AIP conference proceedings, vol 1857, p 030001. https://doi.org/10.1063/1.4987060
97. Scherz JP (1974) Errors in photogrammetry. Photogram Eng
98. Schumm SA (1956) Evolution of drainage systems and slopes in badlands at Perth Amboy, New Gersey. Geol Soc Am Bull 67:597–646

99. Schwendel AC, Milan DJ (2020) Terrestrial structure-from-motion: spatial error analysis of roughness and morphology, vol 350, p 106883

100. Shields A (1936) Anwendung der Ahnlichkeitsmechanik und Turbulenz forshung auf die Geschiebebe wegung. Mittelungen der Pruesspsichen Versuchsanstalt fur Wasserabau und Schiffbau, vol 26

101. Smith SA, Barnard PL (2021) The impacts of the 2015/2016 El Nino on California's sandy beaches. Geomorphology 377: 107583

102. Snavely KN (2011) Scene reconstruction and visualization from internet photo collections: a survey. IPSJ Trans Comput Vis Appl 3:44–66

103. Snavely KN (2008) Scene reconstruction and visualization from internet photo collections. PhD Dissertation, University of Washington, 192p

104. Speitzer G, Tunnicliffe J, Friedrich H (2019) Using structure from motion photogrammetry to assess large wood (LW) accumulations in the field. Geomorphology 346:106851

105. Stockdon HF, Sallenger AH Jr, Holman RA, Howd PA (2007) A simple model for the spatially-variable coastal responses to hurricanes. Mar Geol 28:1–20

106. Stout JC, Belmont P (2014) TerEx Toolbox for semi-automated selection of fluvial terrace and floodplain features from lidar. Earth Sruface Process Landform 39:569–580

107. Strahler AN (1964) Quantitative geomorphology of drainage basins and channel networks. In: Chow VT (ed) Handbook of applied hydrology. McGraw Hill, New York, vol 4, pp 39–76

108. Sunamura T (1992) Geomorphology of rocky coasts. Wiley, Hoboken, 302p

109. Suo C, McGovern E, Gilmer A, Cahalane C (2020) A comparison of high-end methods for topographic modelling of a coastal dune complex. J Coast Conserv 24(47):1–10

110. Thoma DP, Gupta SC, Bauer ME, Kirchoff CE (2005) Airborne laser scanning for riverbank erosion assessment. Remote Sens Environ 95:493–501

111. USGS Kansas Water Science Centre (2020) Fluvial geomorphology. https://www.usgs.gov/centers/kswsc/science/fluvial-geomorphology?qt-science_center_objects=0#qt-science_center_objects. Last accessed Jan 2020

112. Vazquez-Tarrio D, Borgniet L, Liebault F, Recking A (2017) Using UAS optical imagery and SfM photogrammetry to characterize the surface grain size of gravel bars in a braided river (Veneon River, French Alps). Geomorphology 285:94–105

113. Verrelst J, Geerling GW, Sykora KV, Clevers JGPW (2009) Mapping of aggregated floodplain plant communities using image fusion of CASI and LiDAR data. Int J Appl Earth Obs Geoinf 11:83–94

114. Walling DE (1988) Measuring sediment yield from River Basin. In: Lal R (ed) Soil erosion research methods. Soil and Water Conservation Society, pp 39–74

115. Wang C, Yu X, Liang F (2017) A review of bridge scour: mechanism, estimation, monitoring and countermeasures. Nat Hazards 87:1881–1906

116. Wassmer P, Font E, Gomez C, Iskandarsyah TYWM (2020) Magnetic susceptibility and anisotropy of magnetic susceptibility: versatile tools to decipher hydrodynamic characteristics of past tsunamis (Chapter 16). In: Engel M et al (eds) Geological records of tsunamis and other extreme waves, pp 343–363

117. Westoby MJ, Brasington J, Glasser NF, Hambrey MJ, Reynolds JM (2012) 'Structure-from-motion' photogrammetry: a low-cost effective tool for geoscience applications. Geomorphology 179:300–314

118. Westoby M, Lim M, Hogg M, Dunlop L, Pound M, Strzelecki M, Woodward M (2020) Decoding complex erosion responses for the mitigation of coastal rockfall hazards using repeat terrestrial LiDAR. Remote Sens 12(2620):1–22

119. Westoby MJ, Lim M, Hogg M, Pound MJ, Dunlop L, Woodward J (2018) Cost-effective erosion monitoring of coastal cliffs. Coast Eng 138:152–164

120. Wierzbicki G, Ostrowski P, Mazgajski M, Bujakowski F (2013) Using VHR multispectral remote sensing and LIDAR data to determine the geomorphological effects of overbank flow on a floodplain (the Vistula River, Poland). Geomorphology 183:73–81

121. Woodroffe CD (2002) Coasts: form, process and evolution. Cambridge University Press, Cambridge, 623p

122. Woolard JW, Colby JD (2002) Spatial characterization, resolution and volumetric change of coastal dunes using airborne LIDAR: Cape Hatteras, North Carolina. Geomorphology 48:269–287

123. Yang Z, Teller JT (2012) Using LiDAR digital elevation model data to map Lake Agassiz beaches, measure their isostatically-induced gradients, and estimate their ages. Quatern Int 260:32–42

124. Yano A, Shinohara Y, Tsunetaka H, Mizuno H, Kubota T (2019) Distribution of landslides caused by heavy rainfall events and an earthquake in Northen Aso Volcano, Japan from 1955 to 2016. Geomorphology 327:533–541

125. Young AP (2018) Decadal-scale coastal cliff retreat in southern and central California. Geomorphology 300:164–175

126. Young AP, Carilli JE (2019) Global distribution of coastal cliffs. Earth Surf Porcess Landform 44–6:1309–1316

Pointcloud and Hillslope Geomorphology

5

Abstract

This chapter continues the progression upstream, and it then presents the use of pointcloud technologies for hillslopes. The chapter starts with a short overview of some of the slope processes and their numerical expression and how those concepts can be used for the application of geomorphology to hazard management. The chapter then presents a set of applications of pointcloud technologies to hillslope landforms characterized by different scales. As the scales vary, they present different challenges both in term of data acquisition and processing. The slopes are thus divided between interfluves, the drainage network and the mass movements. The chapter ends with a few worked examples using LiDAR and SfM-MVS data, including cleaning the dataset from vegetation, using the vegetation and trees as a proxy of surface processes.

Learning Outcomes: Be Able to Apply the Right Method for Hillslopes

After reading this chapter, the student of geomorphology, who already has a good understanding of hillslope processes, will be able to understand how to generate pointclouds for her/his needs and to do so with the method appropriate to the researched object. The student will also be able to link his/her knowledge to methods and tools in surveying and civil engineering, which are being progressively incorporated in geomorphology as quantification methods, and which, in turn, are in need of the pointcloud technologies. This interplay will show that for good civil engineering calculations, accurate and precise topographic data are also very important.

As the approach to the acquisition of the topographic data will vary depending on the type of landform addressed in the "hillslope environment", you will also be able to relate the major landform geometries and land cover to the appropriate methods and tools to be used.

Finally, you will be able to solve different sets of basic problems related to pointcloud usage for geomorphological purposes using open-source tools and software such as CloudCompare and the R programming environment. The chapter has a set of examples and worked examples I invite you to build on using your own data.

Objectives and Content

The objectives of this chapter are

(1) To provide a very short overview of some of the slope processes and their numerical expression to emphasize the role of HRT pointcloud technologies;
(2) To learn the applications of pointcloud technology to meet the challenges related to hillslope geomorphology;
(3) To be able to use a simple processing method to work on your own dataset.

With these objectives in mind, in this chapter, I invite you to review some of the common concepts of hillslope geomorphology, especially some of the methods used to measure slope stability, as they are contingent on having an accurate topographic dataset. Then, I divided hillslope landforms into different groups (interfluves, drainage network, mass movement deposits and fans, etc.) following a systematic division from the point of view of pointcloud acquisition and processing. This characterization may not follow your typical geomorphological division of space and landforms, but it should help you understand specific challenges related to different types of geometries on the slopes.

Finally, I am finishing the chapter with a set of worked examples at different scales, to illustrate the content of the chapter and also provide you with the opportunity to hone your skills. First, I am giving you an example of road-side cut-slope monitoring using LiDAR and SfM-MVS, with a set of GCPs recorded with a total station. Then, I am giving you an example showing how you can align a point-cloud created by UAV SfM-MVS with an existing ALS dataset,

© Springer Nature Switzerland AG 2022

C. Gomez, *Point Cloud Technologies for Geomorphologists*,
Springer Textbooks in Earth Sciences, Geography and Environment,
https://doi.org/10.1007/978-3-031-10977-5_5

and how known-point usage as GCPs helps improve the accuracy of the SfM-MVS dataset. Finally, I am showing you a method to "detrend" a complex topography to easily extract trees and surface features, without having to work through a complex filtering process. Finally, we are using an example to discuss the potential of using trees and vegetation as a proxy of geomorphological processes.

Who is this chapter for? If you are a student of geomorphology with a good numerical background (rather than social sciences), you can then skip the first section, but if your geomorphological background is more "descriptive", you may then want to read this section and do related reading. The second section is for everyone to read with examples of pointcloud use for different types of landforms. The last section is also for everyone to read.

Who is this chapter not for? This chapter like the rest of the book is at introductory level, so that any specialist of slope geomorphology and pointcloud technology will certainly want to skip the chapter all together, although the examples at the end might give some of us some different ideas.

Introduction

Crozier et al. started their chapter on the stability of hillslopes in New Zealand by "Hillslopes are affected by many types of weathering and erosional processes. Weathering and the removal of rock and soil material in solution occur almost imperceptibly, and even the removal of fine materials by surface wash is seldom observed on hillslopes because, below the vegetation line, nearly all slopes in New Zealand have a virtually continuous vegetation cover. It is only where the cover is broken by gullies or landslides that large masses of debris are carried into valley floors and changes in hillslopes are visible. Landslides are the most obvious and important processes acting on hillslopes in New Zealand. The less obvious processes of soil and rock creep may be important in preparing zones of weakness which will later be exploited during a slide, or in moving material slowly downslope, but the quantitative importance of creep in total denudation is mostly unknown, although assumed to be small" [11].

In the first paragraph of their chapter, Crozier et al. [11] express all the important points of slope geomorphology from a process perspective: (1) the multitude of scales at play, including hardly perceptible processes, (2) the vegetation cover that hides some of the early processes in gullying and also the small-scale processes, and (3) processes that are slow enough to be hard to measure, and it also contains most of the clues to identify correct pointcloud technologies to measure what was thought not "measurable" before. It is worth noting that it is anachronous (as

pointcloud technologies were not developed when Crozier wrote those lines), but his paragraph can be used to emphasize the important role that pointcloud technologies can play in the study of Earth surface processes: (1) "seeing through" the vegetation with LiDAR; (2) being able to repeat surveys using drone photogrammetry and LiDAR; (3) being able to acquire dataset to a few centimetres' accuracy, so that processes that were invisible so far can also be revealed and measured.

Hillslopes represent about 9/10th of the emerged land surface, the rest being dominated by river corridors and floodplains. By comparison, coasts are only a tiny proportion of the surface—although a huge stake for human communities. From this landform distribution, the difficulty to access some of the high-mountains, pointcloud acquisition from airborne and remote stations (with a long-range laser for instance) is therefore essential to the progress of mountain and hillslope Geomorphology.

Because the Earth is a tectonically active planet, where reliefs are constantly created by vertical push of volcanoes, mountain chains, in turn compensated by erosion, collapses and subsidence (at rifts for instance), the Earth is then a planet of gradients. It is thus very logical that the Earth is covered in hillslopes, which take various forms, shapes and gradients, including (a) the interaction of the type of material they are made of, (b) the gradient generated and the weathering process at play, which are themselves mostly controlled by the climate and the weather.

Hillslopes are therefore evolving geomorphologies that can change rapidly under the influence numerous gravity- and weather-driven processes, such as debris flows and landslides for instance. For this purpose, laser scanners and structure from-motion photogrammetry appear to be tools of preference to the geomorphologist. Indeed, if the geomorphologist wants to measure let say erosion rates on a slope from multiple measurements, even with a decade interval between measurement, one will need high-accuracy data and a large number of samples to make a meaningful measurement (and yet, it will only be the result of the ten years' time-span, a different weather, climate, ground cover could yield very different results and even those cutting-edge technologies would not be a silver bullet).

5.1 A Rapid Overview of Hillslope Geomorphology and a Selection of the Processes That Make Them

As for this chapter as well, I believe that the readership will be made of undergraduate and eventually graduate students (and eventually more seasoned researchers and academics who are interested in the topic of point-clouds in geomorphology), and I am thus starting the chapter with a few words on slope processes and landforms. It is however not an exhaustive nor even an extensive introduction to hillslope geomorphology and should be taken as a window on, or a door to the more technical section of the chapter on point-clouds. The choice of one landform over another has just been motivated based on the connections with the later part of the chapter, and one should not feel offended if she or he cannot find his or her favourite landform mentioned in the text.

5.1.1 Focus on Some Remarkable Hillslope Processes

Logically, as a hillslope signifies a vertical gradient, gravity and the temperature gradient and weather modification will be two of the most important ingredients in hillslope processes. And as precipitated water (in liquid form, snow or ice) will then move along the slopes (as water, snow or ice), it is a combination of these that will drive the slope processes as we understand them today. For mountain areas, Evans [16] stated that the six fundamental components of reliefs are the elevation, downslope gradient, across-slope gradient, aspect, vertical and horizontal convexities, which are all derivatives of the surface topography, putting the shape and the landforms at the heart of geomorphology, providing the essential controls over the processes that shape those landforms.

5.1.2 The Role of the Relief and Gravity in Driving Denudation Processes

Hillslope processes occur chiefly under the action of gravity (in collaboration with other processes, notably water born [12]) and their role is to smooth the topographic gradient, notably by transporting downslopes the material in pieces of a broad range of dimensions. This flux has been expressed over long-period of times as a diffusive process, so that for a topography with a z vertical component and a x horizontal component, the material discharge Q_m is defined and with a magnitude of material transport for a given slope k, the material discharge can be calculated as follows:

$$Q_m = -kx\frac{\partial z}{\partial x} \qquad (5.1)$$

In the equation above, k is associated with x, as k will vary with the material and thus in space x. And a general form of long-term slope process can be further simplified to:

$$Q_m = -k\frac{\partial z}{\partial x} \qquad (5.2)$$

Such an equation works for slopes where a soil (in the engineering sense of it) can be generated over the regolith to form a continuous or semi-continuous mantle to be exported as a function of the slope gradient. At a broader scale, empirical relationships relating the denudation rate and the local relief have also been developed. Ahnert [1] found that $D = 0.1535E$ (where D is the denudation rate in mm/1000 years and E is the local relief in m/1000 m). The concepts formulated above are useful when looking at processes such as slow creep, progressive movement due to water transport, etc., which are all a function of the topographic gradient, but when the movement is controlled by subsurface structures, like for landslides for instance, the material transport is not a function of the surface gradient anymore. In such a case, a stable deposit will develop at the base of the slope (it can reach considerable distances) and the upper portion of the slope will be over-steepened. Following such an event, the slope gradient will then tend towards smoothing following the equation above. These two considerations are part of a traditional discussion in geomorphology: i.e. the relative importance of long-lived slowly acting on the hillslope, against low-frequency and high-magnitude impact events.

5.1.3 Hard-Rock Hillslopes

Hillslopes eroded and cut into hard-rocks can sustain very steep slopes, as it can be seen for Fjords in Norway, to the south of the South Island of New Zealand or in South-America for instance. From an endogenous processes perspective, slopes cut in hard-rocks are explained by sustained tectonic uplift, where erosion dominates over deposition. From a hydroclimatic perspective, they can also be linked to glacial and periglacial domains where the right conditions for the development of soils are not met. The rock mass can be sheeted, fractured, etc., and this characteristic creates discontinuity in the shear strength, which, in the form of rock joints notably, have different incidence on the erodibility depending on their location and orientation with respect to the slope (e.g. a fracture parallel or one that is perpendicular will have different effects). Furthermore, the type of rock is another essential control factor (whether it is igneous, sedimentary or metamorphic), and one can differentiate them based on the mineral content, the texture of the rock and the fabric. In a classic text now, Patton [48] researched the relation between joint planes irregularities and the shearing strength, which resulted in the shear strength along the rock joint (S), as follows:

$$S = c + \sigma_n \tan(\varphi_r + i) \qquad (5.3)$$

where

S: shear strength at the joint,
c: cohesion of the rock joint,
σ_n: normal stress,
φ_r: the residual angle of friction of the rock surface,
i: a measure of the roughness angle.

The structure of this equation is very similar to the one used for sediment slopes, which often define the shear strength of soils from the Coulomb's theory, but Eq. 5.3 is applied to rock masses with a discontinuity along the sliding direction. A comparable concept was also applied by Barton and Choubey [6] for instance:

$$S = \sigma_n \tan\left[\varphi_b + \text{JRC} * \log_{10}\left(\frac{\text{JCS}}{\sigma_n}\right)\right] \qquad (5.4)$$

where

JCS: compressive strength of joints,
JRC: roughness coefficient of joints,
S: shear strength along the joint,
σ_n: normal stress,
φ_b: frictional sliding angle.

5.1.4 Hillslopes in Sediment Formation and Sedimentary "Soft-Rocks"

Hillslope that are made of sediments and soils (the reader will be careful with the use of those two words as geomorphologists, sedimentologists, engineers and soil scientists all use the words with different definitions, but in the present case, I just mean material that can be potentially loose and decompose to a grain level, by opposition to the rock-slopes) or mantled with any loose formations differ from rock-slopes because of the gas exchanges that can exist within the soil and the potential for water to travel vertically and horizontally in the slope at greater speeds than in rockslopes, which are more "monolithic". Furthermore, the movement that is strongly controlled by the rock joints and fractures on rock-slopes is rather controlled by some forms of granular and plastic and visco-plastic relations. Part of the material can even be turned into liquid flows.

The Role of the Hydroclimatic System in Controlling the Processes

After the topography and its derivatives, the second set of factors that control hillslope processes is the hydroclimatic system. The latter plays a role above the slope (atmosphere) and also within the slope (groundwater movement, freezing, tunnelling …), generating a very broad range of processes and landforms. In this short-hand, we are only addressing the

former one, from the angle of water (in its different phase) above the surface. Depending on the temperature, water will either be precipitated and travel over the topography as liquid water, snow or ice, and the total amount and the amount of water over time (intensity of precipitation, etc.) will also help defining how the water is transferred downslope and how it carries material (torrents, mud and debris flows, landslides). Once again, we are dividing different units of processes and related landforms, but the reader will want to understand them as continuum, rather than individual blocks, but it is easier to think about a typical glaciated and paraglacial landscape for instance and how point-cloud technologies can be used in one or another situation. In the field however, there will be a wide range of shades between the different components of the landscape.

Glaciers as Geomorphic Agents

When the temperature balance is negative or around 0, snow will accumulate and eventually change into ice and flow under gravity. "A glacier is a natural accumulation of ice that is in motion due to its own weight and the slope of its surface. The ice is derived from snow, which slowly loses porosity to approach a density of pure ice" [3, p. 214]. Although the glacier starts in an accumulation zone (cold or with a lot of snow precipitation to compensate for the loss), the lower part of the glacier is the ablation area, where the glacier "melts". Therefore, although the front might not be moving spatially, the glacier keeps flowing. In locations of the world where heavy precipitation feeds the glacier in the accumulation area, the lower tongue can extend to low altitude (e.g. at Fox Glacier on the West Coast of the South Island of New Zealand [53, 54]).

Glaciers are an essential to understand mountain landforms worldwide, even if their extension and number is dwindling under the effect of anthropogenic climate change (after temporary extension in the Southern hemisphere due to increased precipitation). They have carved valleys and deposited a series of moraines and debris through the Quaternary period, and this, due to two main mechanical erosion processes: glacial erosion and glacial hydrology erosion. Glacial erosion can be divided between glacial abrasion and glacial quarrying, which fracture the bedrock material and entrain it in the ice. Glacial hydrology with meltwater plays a role downstream the glacier as well as it does underneath the glacier, all contributing to glacial landforms [8].

Landslides and Debris Flows

Landslides are the movement of rock, soils, bio-materials and anything covering the moving surface, under the action of gravity and auxiliary processes. Because of their spectacular aspect in the landscape, and the needs to assess hazards and associated vulnerabilities, landslides have attracted a large amount of work in fundamental research

[13] as well as in the field of disaster risk [27]. This interest has kept momentum, because the processes are only approximated, and also because climate change is modifying existing dynamics and processes [39], with for instance the relative-elevation of the water base level for coastal landslides, increased rainfalls and modification of the temperature balance in mountainous areas, etc. Landslides can be divided following the two different types of material we have defined above for the landforms: rocky material or soil, producing different types of slope movements. On hillslopes, the common types of landslides are (1) the translational slide, which occur in coherent material, moving as discrete failure often along existing stratigraphic plane weaknesses, (2) the rotational slide, which moves along a convex surface, most often as a coherent or semi-coherent block of plastic material, (3) the topples and falls that are movement from sub-vertical surfaces, often in the form of discrete groups of blocks, (4) rock avalanches or Sturtzstroms, which are large movement of rocks and debris with a high mobility and velocity, and in which the material becomes fragmented due to the movement, and (5) flows such as debris flow and mudflows which are mixtures of water and sediments, clasts and debris flowing in channels.

Landslides and debris flows are engaged in a multi-layer complex relation (like an old couple maybe). Taking the example of Ohya-Kuzure Landslides in Shizuoka Prefecture (Japan), a 120 million m^3 mass movement occurred in 1707 [64], which destabilized most of the watershed it is located in. The displaced and deformed sandstones and mudstones are presently moving under the effect of gravity, weathering under heavy rainfalls [30] and freeze–thaw, in such a way that debris flows are being fed by the material made available by present activity, but "prepared" by the eighteenth century mass movement. In combination with the rainfall patterns, the material availability then defines initiation and runout of the debris flows [28].

Considering a broad classification, debris flows are part of the landslide phenomena, but from a point-cloud acquisition technology by laser scanner or by photogrammetry in the field, the two phenomena can be separated as most landslides will allow for several data acquisition, representing several time steps in the movement, while debris flows are more complex to capture during their travel.

5.1.5   Applied Hillslope Geomorphology and Civil Engineering

Hillslope is maybe the topic where civil engineering and geomorphologists meet and influence one another the most (in my humble view), and one of the concept that has been influencing geomorphology are the models of slope stability (e.g. [33, 36, 40, 58, 69]). Although there are plenty that

have been developed for engineering purposes—all potentially stemming from the important work of von Terzaghi [60]—the two main equations that are being extensively used are the infinite slope model and the XXX model, formulated in the factor of safety. In those models, engineers and geomorphologists consider the soil as a continuum and they apply geometrical relations to balance the forces that initiate movement against those that hamper it.

The infinite slope model is well adapted when the soil layers are parallel to the topography. Therefore, it is often used on slopes covered by volcanic ashfalls for instance or when the slope formations have been created or deposited in close to parallel layers to the surface. This relation is often expressed as:

$$F = frac C + \{\gamma_t(D - H) + H(\gamma_{sat} - \gamma_w)\} \cos^2 \beta \tan \varphi \{\gamma_t(D - H)\} \cos \beta \sin \beta \tag{5.5}$$

where

F is the factor of safety,
C is the cohesion,
φ is the angle of internal friction,
γ_{sat} is the saturated unit weight of soil,
γ_t is the wet unit weight of soil,
γ_w is the unit weight of water,
D is the soil thickness,
H is the water depth below the surface,
and, β is the slope angle.

Another popular approach is to consider the sliding surface, not as a straight line but a loop, which is part of an arc, being a common sliding surface on hillslopes. By separating the slide into a finite number of vertical slices, the factor of safety F can then be expressed by a few methods, such as the Janbu method, the Bishop method, the Morgenstern–Price method, the method of slice, and from charts' methods (e.g. the Spencer chart, the Taylor chart, etc. …). As an example a simplified formulation of the method of slice for an isotropic earthquake loading on a slope would be expressed as:

$$F = \frac{\sum_{i=1}^n \left(C_i \frac{b_i}{\cos \theta_i} + \left(W_i \cos \theta_i - \alpha_i u_{wi} \frac{b_i}{\cos \theta_i} \right) \tan \varphi_i \right)}{\sum_{i=1}^n W_i \sin \theta_i} \tag{5.6}$$

where the terms that were not define in the infinite slope equation are:

b: the width of each slice,
α: the area ratio for the water phase,
u_w: the water phase,
W: the total weight of the slice,
and i and n are subscripts that refer to the ith slice in a total population of n-slices.

In both cases, F is therefore a measure of whether the slope is prone to sliding or not.

One can see how those formulations can be useful for slope stability and the classification of slopes. Even if the equation would suggest that 1 is the hinge value in the F formula, engineer often use values such as 1.2 and 1.25, to leave some wiggle room to account for the error margins in measurement and calculation. The formulations above are usually set for loose formations, soil-mantled slopes, and still based on geometric approaches, those concepts have been also applied for highway roadsides evaluation of rockfalls for instance, using semi-quantitative methods. In USA, the Federal Highway Administration has developed a set of field methods to score the hazard levels of rockfalls [50, 51]. This method combines a geographic database of rockfalls along the highway, a rating of slope to identify priority areas, and project identification, development and costing, as well as a review system of the rockfall hazards. This is the typical type of methodology that can be improved with the help of point-cloud technologies for a whole highway and to review those measures every year (not feasible in turn of human resource costs necessary). In other words, ALS and vehicle-based LiDAR can monitor highways at high resolution, by just driving by them.

Although slope processes encompass numerous other processes on top of the ones chosen above, this choice was motivated by (1) realizing how fast hillslope can change and how the pointcloud generated by geomorphologists can come and help civil engineers and resource engineers.

5.2 Hillslope Geomorphology and Point-Cloud Technology

The processes and associated methods presented above are all contingent to having an accurate slope geometry, and that's when pointcloud technologies become superior to other methods, providing that you understand the limitations and the pitfalls of the different methods. Moreover, hillslopes are also defined by a variety of "sub-forms" and we will divide them based on the specific challenges to pointcloud technologies. I have thus chosen to divide hillslopes into three groups of catena differentiated as follows: (a) catenas associated with the geomorphology of concentrated drainage; (b) those associated with the interfluves; and finally (c) those associated with the contacts of the hillslopes with the surrounding land.

5.2.1   Landforms Associated with Hillslope Drainages

Difficulties	– The walls of the drainages can be steep and difficult to capture – Water-level change can have important impacts on – The low position of the drainage can be blocked by a dense vegetation cover
Recommended tools (not exhaustive)	– Helicopter-based LiDAR for slow movement – UAV-based LiDAR – Ground-based slam and hand-held scanners

Hillslope Streams, Thalwegs and Gullies

Arguably, gullies, thalwegs and other drainage systems on the hillslopes could have been integrated to the chapter on rivers, but we separated the former from this type of drainage based on the gradient and the type of process that carve them in the landscape, even if they both look "similar" in morphology. This may seem a very geomorphological intent, but it also has technical implications for pointcloud acquisition. For the hillslope drainage networks, the comparatively high slope gradients can favour debris flows and other torrential flows, which result in landforms that are different from those found in floodplains and along fluvial corridors in detail, and this gradient can be expressed as the stream length gradient index (SL) [22, 56]:

$$\text{SL} = \frac{\Delta h}{\Delta l} * L \qquad (5.7)$$

where h is the elevation in the direction of the stream segment, l is the length of a segment and L is the total length of the stream from the divide to the middle section of the calculated segment. Once again, this gradient is only as accurate and precise as the Gullies and channel upward limit is the channel head, which is the highest point where water and sediment flow starts to concentrate [43]. Plackowska et al. [52] combined TLS with point densities between 1 and 2500 pts/m^2 and ALS at 1 pt/m^2 to calculate 16 headwater characteristics of forested channel heads in the Polish Carpathians. Their work shows that to estimate the position, and geometric characteristics of the channel heads, even ALS underestimate the size and position (providing that we are interested in sub-metre resolution). Because TLS is often an expensive equipment to buy and thus deploy, there is work on the channel heads that have been using TLS as a benchmark and then the following 3D data acquisition and construction was performed using photogrammetry for instance. In Spain, Gomez-Gutierrez et al. [20] have used a Leica C10 Scan-station (± 2 mm error at 50 m distance from the TLS) combined with then repeated photographic sets (SLR Canon 550 D). For five channel heads, they collected between 41 and 93 photographs per location, producing photogrammetrically derived point-clouds with 297–901 k points, while the TLS provided values between 531 and 3181 k points, with point densities for the two techniques between respectively comprised between 7913 and 17,973 pts/m^2 and 13,994 and 47,598 pts/m^2. This dataset shows the superiority of close-range acquisition procedures when working with small-scale features and especially on slopes, where the exposition to an inflight laser scanner is not optimal. The issue of using the ALS for gully and channel heads is further acute for high and steep slopes. At Ohya-kuzure, in Japan, Imaizumi combined ALS from 2003 to 2007 as well as TLS data of 2010 and 2011 [29] to measure the retreat rates of the sub-vertical heads, and he calculated a maximum local error of 30–40 cm vertically for the ALS data. After processing the topographic data to limit the error margins (see [29]), the authors could deduce erosion rates ranging from 0.51 to 0.22 m/year and they also had measurement that were dense enough to consider the dip angles of the alternating sandstones and mudstones at the site. This second point is certainly one of the new possibilities that point-cloud technologies are opening to the geomorphologist, the possibility to define the relation between rock and sedimentary formations and the landforms. A relation that was long "assumed", or deduced, as it could not be measured can now be evidenced from the data. The reader will note that the problem of data acquisition for sub-vertical surfaces extends more largely to any sub-vertical surface, such as rock walls (i.e. [14] who used TLS to obtain the geometry of a sub-vertical wall in the Pyrenees Mountains).

Leaving the gully and channel heads, for their main stems, the change over time of the gully vertical and horizontal geometry from chronological ALS datasets [5, 15, 61] and from TLS [21] has been used to calculate erosion, soil losses.

Knickpoints

Difficulties	– In bedrock, the knickpoint can move very slowly – Knickpoints associated with cascades experience constant geometric change with the water – Cascade water can hide and hamper measurements
Recommended tools (not exhaustive)	– TLS – SfM-MVS combined with TLS – UAV LiDAR

Knickpoints exist in fluvial areas, but their density and amplitude increases in hillslope, as the gullies and the streams tend to encounter deformed, faulted and varied lithologies (not that such setting does not exist in fluvial landscape, but it is arguably more representative of hillslope landscapes). And the knickpoints are then often associated with waterfalls.

If you think about the erosion on hillslopes as a system with boundary conditions, the downstream boundary condition in space for a slope is located at the local gully or channel it connects to. Therefore, one important control on this boundary condition will be the knickpoints. This same knickpoint will in turn control the sediment transport in the gully or channel.

Because knickpoints tend to move at a low pace (e.g. Ito-Fudo falls at 0.0013 m/year to Soho falls at 0.13 m/year in the Boso Peninsula according to Hayakawa and Matsukura [26]), point-cloud technologies from LiDAR notably have been particularly instrumental in constructing an accurate image of the retreating rates.

Using a LiDAR-based DEM of North Carolina (USA) generated in 2006, Gallen et al. [18] extracted the longitudinal profiles of the Cullasaja River from which they derived metrics for three main knickpoints, and they linked the change of profile in between the knickpoints and the tributaries' profiles to the development of hillslopes.

In Japan, Hayakawa and Oguchi [25] have demonstrated that the distribution of knickpoints was controlled by relative altitude, with a higher density in the upstream sections of the channels, and so regardless of the lithology. They concluded that for steep-sloped channels, the linear density of knickpoints in Japan is 0.5 per km. They further explained that the presence of knickpoints on Quaternary volcanic hillslope seems to be controlled by the volcanic activity. Combined with dating of the rock formations, it is then possible to measure the retreat rates of the knickpoints. Hayakawa and Matsukura [26] have empirically determined that waterfalls (i.e. knickpoints) cut in Pleistocene, Pliocene and Miocene material were retreating following this equation (where the density of water was replaced by 1 here):

$$\frac{D}{T} = 99.7 \left[\frac{AP}{WH} \sqrt{\frac{1}{Sc}} \right]^{0.73} \tag{5.8}$$

where D is the retreat distance along the channel, and T is the time component of the recession, A is the drainage area above the waterfall, P is the annual precipitation in the catchment and the product WH (Width * Height) is the surface area of the waterfall face.

Measuring the change in the position of the waterfall face or the knickpoint along the central axis of the drainage network over several years to decades should therefore provide information on present retreat rates, to notably link those retreat rates to extreme events such as debris flow, which tend to impact the morphology of steep channels [65]. Indeed, long-term changes measured from significant modification do not provide a distinction between what is the role of debris flow against traditional stream-flows for instance. Being able to measure those changes (only a few millimetres to centimetres for a couple of years) with laser technology for present events does provide researchers with new insights on the evolution of knickpoint. We don't have to wait a couple of million years to extract a rate from an averaged dataset (which will encompass a lot of processes we may just have no record of).

Glacial, Pro- and Periglacial Landforms and Point-Cloud Technologies

Difficulties	– An ever-moving landscape
	– Debris-covered slopes are all unstable
	– Often narrow and deep valley offer limited sky-view
	– Steep topographies are hard to address from the sky
Recommended tools (not exhaustive)	– TLS
	– UAV LiDAR

When we drop the temperature, flowing water is replaced by flowing ice, generating on hillslopes with the right climatic and slope conditions, alpine glaciers for instance. For glacier monitoring and mass balance determination, LiDAR has revealed to be a valuable tool ([4, 49]), notably for the construction of DSMs in glaciology ([45, 17]). The action of glaciers in the landscape create typical landforms, one of which is certainly the cirque, where the glacier is starting from. Although the reason why a cirque and a glacier develops in one place or another is still an egg and chicken debate in geomorphology (do we have a notch because a glacier carved it or did the glacier started at a given location because a notch existed), there are no doubt about the important of the erosion processes occurring in them [7]. Hartmeyer et al. [23] have used a Riegl LMS-Z620i laser scanner and a high-resolution camera to quantify the evolution of the cirque walls by 56 rockfall events over a 6 years' period. They could measure that the eroded rockwall volume was 2551.4 ± 136.7 m^3 corresponding to a 1.86 mm/year (± 0.1 mm) of wall retreat. Because the retreat is not linear everywhere, but highly dependent on the rockfall activity, the authors also noted that the retreat is highly variable. This is where the power of point-cloud technology offering dense measurement in both space and time is showing its advantage over other methods. Furthermore, rockwalls being a sub-vertical feature, traditional 2D DEM or DSM

representation of the data (in a GIS system for instance) would not be adequate for this work.

In these environments, one can also find a broad variety of secondary landforms, which are often made of coarse and angular material that makes them difficult to survey using traditional methods. These landforms also combine with one another, across different spatial scales, and only pointcloud and pointcloud-derived data provide the right tool to address this diversity. In the Troutbeck Valley (Cumbria, UK) for instance, Miller et al. [42] have mapped inherited glacial geomorphology made of drumlins, moraine and moraine benches, demonstrating the "one dataset fits all" (scales) approach that is possible with HRT point-cloud technologies. When glaciers have flown over lower-gradient topographies, ribbed moraines which are morphologically hummocky formations [62] have been created from subglacial erosion. Using a Leica ALS60 laser scanner flown between 1800 and 2445 m altitude, Middleton et al. [41] have mapped the geomorphology of such ribbed moraines in Finland. As the HRT data provides sufficient details on the morphology of the moraines and for instance drumlin fields [59], it is then possible to also use the ALS data to reconstruct the flow paths of past glaciers [46]. Pointclouds are thus giving you the opportunity to work on present and inherited landforms as well as the processes that lead to them.

5.2.2 Landforms Associated with Hillslope Interfluves

Difficulties	– Large spatial extent with often relative limited local variability
	– Potential dense vegetation
	– Potentially hard to access and ground control points can be difficult to place on the slope
Recommended tools (not exhaustive)	– TLS if the slope is small enough of steep enough to be seen from sets of vantage points
	– UAV LiDAR
	– ALS

Processes occurring in between the "veins" of the drainage network are an essential component to controlling the erosion rates in the drainage network and in the erosion processes of hillslopes. Traditional point-measurement methods are insufficient to provide an accurate view of erosion processes, and even their combination with more recent radionuclide dating of surface exhumation and erosion for instance necessitates averaging over long periods of time and are also tributaries of dating models. HRT point-cloud technology however provides an accuracy between a few to a few tens of millimetres using TLS, in such a way that erosion can be "measured" and spatial variability can also be revealed.

Furthermore, single-point methods (both topographic and radionuclide) are potentially representative of a relatively small area around them, but once the interfluve shows large extent, not only the erosion variability increase, but so is the number of temporary storage areas, which are often flushed during heavy rainfall and seismic events. Without a good understanding of these temporal and spatial variations, the erosion rates assessed from stream transport for instance are difficult to understand correctly. This is particularly true for mountainous hillslopes that were glaciated during the Quaternary, as huge stocks of material are still on the interfluves, and understanding their evolution and predicting their change is one of the challenges remaining in geomorphology. In Arizona (USA), Wawrzyniec et al. [66] have worked from two sets of TLS data they collected in Badlands, and in the period June–October 2006, the authors measured the smoothing and erosion of different features that were the result of hillslope erosion. This method worked well with TLS because the examined features were local and with virtually no vegetation.

5.2.3 Tectonic Deformations and Ruptures

Difficulties	– Deformation with a wall can be difficult to measure from aerial position in NADIR – Ruptures tend to be smoothed out rapidly and rapid assessment can be essential in specific environments – Large-scale deformations, like undulation call for wide-scale measures
Recommended tools (not exhaustive)	– ALS – Airplane and UAV photographs for photogrammetry – TLS for local rupture walls

Tectonic geomorphology investigates the role of tectonic activity in deforming (and forming) mountains, slopes and how those are deformed and ruptured has made immense strides thanks to point-cloud technologies. For instance, exhumed fault scarps are important indicators of shallow earthquake present and past activity (e.g. [55, 67]), and ALS and TLS have been intensively applied during the last 10–20 years in order to measure the offsets and movement that resulted in those scarps [31, 32, 34, 37, 68]. Although TLS provides the best results on sub-vertical (yet local) features, large-scale fault-line scarps extending several tens of kilometres, yet with an offset of only a few tens of centimetre to a few metres can be imaged with high accuracy and precision using ALS systems. At such scale, as the landscape is most likely to include different land covers and eventually human settlements, ALS provides the right dataset to filter surface features and extract the geomorphology.

In the Carpathian Mountain, LiDAR-based DEM, complemented with fieldwork for mapping the post Last-Glacial-Maximum fault movements near alluvial fans, was successfully used over a rupture more than 50 km long. The calculation of a slipping rate of 0.1–0.3 mm/year (very rapid) has from this method been determined [47]. Because the definition of faults (and geomorphology in general) is a geometric science, point-cloud technologies have allowed the geomorphologist to gather the morphology and morphometry at different scales of these features.

If you are working on visible fault movements or after an event, I can only suggest that you combine multiple tools, with ALS to have an overview of the full event, and chosen TLS or ground SfM-MVS data acquisition of local features, like a wall generated by the fault movement for instance. SfM-MVS can then be replaced by TLS would the vegetation hamper the visual acquisition of the geometry of the wall.

5.2.4 Mass Movements, Slope Collapses and Fans

Difficulties	– Large-scale events can move slowly – Mass movements often occur under vegetation cover – Heavy rainfall events and earthquake can trigger thousands of events – Topographic variations on large fans can be minute to measure
Recommended tools (not exhaustive)	– UAV LiDAR – TLS – UAV SfM

Mass movements have been preferential geomorphological objects for the use of TLS and ALS as well as photogrammetry [57], and the HRT has provided the possibility to map relative movements within one mass movement, even under thick vegetation cover [9]. In such cases, authors have multiplied collection campaigns combining the data and collecting ALS data at different periods of the year (notably during the no-leaves period) [9]. Comparing ALS data of 2010 and 2013, Anderson et al. [2] mapped and characterized 120 rain-triggered shallow mass movements and debris flows (out of the 1100 recorded) using DoD in the Front Range of Colorado. Once again, it is the broad-surface covered with high-spatial resolution both vertically and horizontally that allows the mapping of change <1 m vertically.

For geomorphic and hazard management purposes, the bulk volume of sliding masses is also a parameter that scientists and engineers have tried to extract using HRT point-cloud technology (e.g. [10]), especially when landslides occur by the thousands following heavy rainfall and landslides (e.g. [19, 63]).

As for other landforms and processes, one of the present research topics in pointcloud for landslide research is the development of filtering and processing techniques, as well as methods to compare pointclouds with one another. For two landslides in the Schmirn valley (Austria), Mayr et al. [38] have researched an automated classification two-step method with first a supervised morphometric classification and then a filtering using topological rules. Such automation is essential when there are numerous events to classify and also to compare multiple point-clouds over time and assess potential movement.

At the hinge between hillslopes and the lower land or with a water body, one can find aprons, fans and cones that are the results of material moving from the hillslope as concentrated flows or as open-slope movement, which suddenly "encounter" a slope gradient change and a change of direction in the vector of movement. As the material transported by rockfalls, landslides, debris flows and torrents is a function of the local slope and gravity, the sudden slope angle reduction translates in an accumulation of material at the hinge between the hillslopes and the lower surrounding land, and they are ubiquitous landforms, that can be found in all climatic environment [24].

5.3 Worked Examples

5.3.1 Road-Side Slope Monitoring Using Low-Cost SLAM-LiDAR Sensor and SfM-MVS

Mountain road-side stability is an issue that is ubiquitous to hillslope worldwide, and for numerous communities and professionals with limited human and financial resources, regular TLS or ALS is not always possible. If you are an undergraduate or a graduate student, this book was designed for, you may not have access to a TLS at your university, but you certainly have a camera and your department may be able to afford a low-cost solid-state LiDAR (<900 US$) or even an iPad-Pro with a LiDAR sensor (although you want to make sure that you check and calibrate a sensor that was not designed for surveying or for topographic or for geomorphologic purposes).

At the back of my home in Kobe, Japan, rises the Rokko Mountain, with a lot of roads and paths mostly dug from the hills for maintenance and hazard management purposes. Along one of those roads, one of the corners is cut in fully weathered granite, crumbling as coarse sand after rainy events. I am using this road cut to provide an example on how SfM can be used to create a model of a small hillslope portion, and how it can also be combined with a dataset created from a low-cost LiDAR sensor.

In this example, I have used the iPad-Pro LiDAR sensor with the pronoPointCloud application, which simply returns the points with the X, Y, Z, R, G, B values. The application offers two possibilities to capture the points, one as a snapshot, and one as a series of points that are captured progressively as the iPad is moved and rotated. However, limitations in the accuracy of the accelerometer and the gyrometer tend to create millimetre to centimetre-scale error. For the present application, I therefore took a set of point-clouds which overlaps and I matched them with one another in Cloud Compare, using first the "2 point clouds alignment tool from defined sets of points" and then the "refine roughly aligned point-cloud tool".

Once aligned, there is a scale issue to be resolved for the next step of combining the LiDAR data with SfM-MVS. From the iPad application, the point-cloud is in millimetre; we therefore need to change the scale from millimetre to metre. It is a very straightforward conversion, but if you are thinking about Excel or your favourite equivalent, it is most unlikely that it will be able to handle the point-cloud. I am showing in the box underneath, how you can do it simply and rapidly in R using the LidR package.

```
> library(lidR)

> las<-readLAS(`PCL1.las')

# these three lines modify the scale in X,Y,Z from
millimetre to metres.

> las@data$X <- las@data$X / 1000
> las@data$Y <- las@data$Y / 1000
> las@data$Z <- las@data$Z / 1000


# this line writes the changed data as a new pointcloud
named PCL2

> writeLAS(las, `PCL2.las', index=FALSE)
```

Looking at the two dataset in cloud compare (Fig. 5.1—pay attention to the scale), the right panel shows the results from the few lines of codes above. Once you have a point-cloud in metre, you can use the Metashape(R) software to import the point-cloud, and from this point cloud, you can choose a set of tie points that you will use to calibrate and scale the SfM-MVS point-cloud.

If you wanted to merge two point-clouds that have been built from a laser scanner, the easiest solution would be to

Fig. 5.1 Change of scale from millimetre to metres, screenshot from the display in CloudCompare. The point-cloud contains 1.567 million points

use Cloud Compare because the two point-clouds do not have "deformations" like a potential dome effect that can be generated with a camera-based method for instance. With SfM-MVS, the control points (GCPs) are also helping the process in constraining the position of the calculated points. It is therefore important to add as many points as you can within the SfM process (the first step in SfM-MVS) instead of constructing your SfM-MVS point-cloud first, and then attempting to align it with the LiDAR fsensor constructed point-cloud. If you had for instance GNSS-RTK or PPK data or a dataset from a total station, you could use these points to construct your SfM-MVS point-cloud and then you could compare it with the LiDAR data, but in the present example we are working solely from SfM-MVS and low-cost LiDAR.

The workflow in Metashape-Pro is rather straightforward, from File > Import > points, you import your 3D laser point-cloud, and in the same "chunk", you open your photographs (Workflow, add photographs or a folder with the photographs). Using the tie points tools ("flags" in the software), you pick your points on the 3D pointcloud by just clicking on them, and then you open your photographs (one by one) and click the control points that you have added on your 3D point-cloud to your photographs. One way to work faster with this step, is to print out targets that Metashape can automatically recognize and place them in the scene to be modelled before even do your laser-scan (the targets to be printed can be found in the menu Tools > Target). This way, you can search automatically for the targets in the photographs, and then place those points from the

photograph on your 3D laser point-cloud in the Metashape-Pro software. Once you have your tie points, you can then recreate your surface using the traditional procedure. If the process went "well" your two point-clouds should show little discrepancy when matched against one another. To do this test, you can use the M3C2 algorithm in CloudCompare and compute the difference between the two point-clouds (Fig. 5.2). The result of the experiment shows for a small erosive-surface where tree roots and the base of a tree stem is visible, the error between the two point-cloud is almost at 0 for the majority of the surface. However, there is localized error in the shadow of the tree and surrounding roots where small cavities are present underneath. Using this technique, you can estimate where the error is most likely to be present when you do comparisons at different dates for instance. If your interest is to investigate the quality of a SfM-MVS point-cloud that you are creating for instance, I suggest that you take series of photograph-sets and that you run each set several times with Metashape-Pro, you will see emerging the error due to the SfM-MVS process and you will also see emerging the error due to the number and photograph angles … etc. Even if you cannot quantify the source of the error with this method, you can however define a statistical sets of error margins for the same object, the same camera, isolating angle of view, distance and processing. But as for many processes in sciences (physics or engineering), there is a portion of "art and craft" that can only be acquired by training, which I would advise you to do as much as possible.

Fig. 5.2 Comparison of the 3D model generated using SfM-MVS and the model generated with the 3D LiDAR sensor on the iPad Pro, using the M3C2 algorithm and data collected on the mountain slopes near Kobe City in Japan [35]

5.3.2 Aligning a Debris-Flow Fan Lidar-Based DEM and a SfM-MVS Point-Cloud

In this example, I am showing the importance of "re-processing" SfM-MVS data, in vegetated areas, and when only a DEM (yet a high-resolution DEM) is available. The location is at Mount Yakedake, in Japan for a research project in collaboration with Dr. Miyata of Kyoto University, and the research data has been collected from a site instrumented by the university. In the present example, there is a 1 m horizontal resolution DEM that is available derived from LiDAR data, and aerial photographs I took from a low-cost commercial UAV mounted with a GNSS. Because sediment deposition and erosion has occurred in the valley and on the debris-flow fan, it isn't suitable to pick control points from the LiDAR data at those locations, that's why I chose the road and the edges of check dams. After creating the first model in SfM-MVS (Original SfM data), a set of 213 sampling points from both the original SfM model and the LiDAR shows an important difference (Fig. 5.3). Because the DEM data is an interpolation of the LiDAR data, and because we don't know where the laser touched the ground, it is important to use data that show similarities in Z. For this reason, I suggest using the central part of the road. For this second part of the process, you should import in a GIS software the generated orthophotograph. Placing this orthophotograph over the LiDAR-based.

example uses the slope from a valley at the foot of Mount Rainier in the USA (Fig. 5.4).

```
> library(lidR)

> las<-readLAS
(`WA_MountRanier_2007-2008_000001.las')
> print(las)


class      : LAS (v1.0 format 1)
memory   : 214.1 Mb
extent     : 582360.4, 582750, 5185500, 5186250
(xmin, xmax, ymin, ymax)
coord. ref.: NA
area        : 287891.3 units²
points     : 2.81 million points
density   : 9.75 points/units²


> dem <- grid_terrain(las, 1, knnidw())
> plot(dem, col = gray.colors(50, 0, 1))
> normlaspt <- las - dem
> plot(normlaspt, size = 4, bg = `white')
```

5.3.3 Detrending the Topography to Extract Small-Scale Features and Trees in R

In this example and the following one, I am using one of my favourite "go to library" for Lidar processing, the lidR library, which has a function to detrend a pointcloud. The

5.3.4 Using the Trees as a Proxy of Slope and Other Geomorphological Processes

When I was a student at Paris 1 Sorbonne University, and with the students of my promotion, we were brought on a

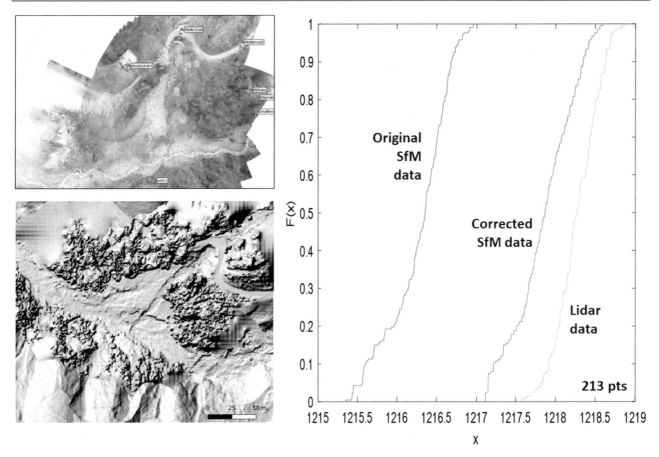

Fig. 5.3 Comparison of the aerial Lidar data against the SfM and corrected SfM for an active fan under thick vegetation cover in the Iida mountains region of the Japanese Alps. The original SfM data was generated from the UAV data and the UAV GNSS device and the on-board accelerometer, without using any ground control points (GCPs). The corrected SfM data used GCPs extracted from the LiDAR data from features that have not moved, between the LiDAR data recording and the drone imagery, such as the edges of check dams, roads … etc. The comparison was made out of 213 identified points. The central part of the image was affected by a debris flow and although it was free of vegetation, geomorphological change did not allow for the use of its surface as a comparison place

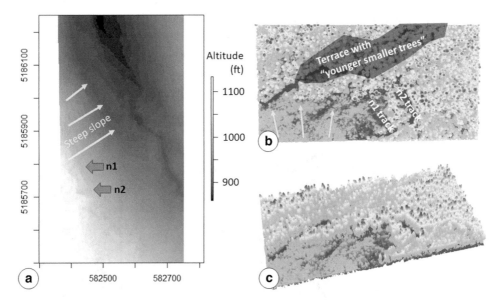

Fig. 5.4 Lidar tile collected in 2008–2008 at Mt. Rainier USA (*Data Source* USGS) showing **a** the gridded topography at 1 m horizontal resolution using the classified point-cloud *n*1 and *n*2 are for two topographic notches on the steep slopes; **b** the pointcloud normalized by the topography, so that the remaining height is only the height of the trees. *n*1 track and *n*2 track are deposition zones where vegetation was either removed or did not grow due to the material movement starting from notches *n*1 and *n*2; **c** is a side 3D view of **b**, where you can clearly see the low vegetation on the steep slopes

one-week tour of the French Alps, and on investigating the hierarchy of processes in the landscape, I remember that tree size and height was used as a proxy to divide the landforms into relative age groups. Today, with the advent of point-cloud technologies for trees and vegetation [44], I would have added that the shape and position of tree branches and tree roots distribution provide further indication of the interaction of trees with Earth surface processes. I am even convinced that there is a much larger field of research to be developed with trees as bio-sensors.

In this example, I am walking the reader through a simple procedure to extract tree height as a proxy of tree age, and thus a relative age of the landform they are built on. It does not mean that the landform the tree has grown into has not seen any form of reworking, but we know that the surface the tree has grown on is unlikely to have been eroded. As tree roots usually "live" in the first metre beneath the surface, it also gives us an indication of the maximum erosion that can occur over a short time-span without killing the trees. The approach is not going to replace fieldwork, but it can be used to define a sampling strategy for dendrochronology for instance, or on the contrary to extrapolate samples into a broader homogeneous population of trees.

For the present example, you need to have finished the one above, because you will need to use the detrended topography, in order to extract the tree height more effectively. I am showing you the results before and after detrending. You will note that the measure including the topography has numerous errors, but although the results from detrended-tree heights provide a better result, there are still errors at the limit and also due to the sampling area to recognize the trees. I can only strongly encourage that (1) you try with different parameters that you can determine empirically by investigating your data manually first, and (2) you compare the automated samples with a few plots where you have done the work by hand, so that you get an idea of the error induced by the algorithm (Fig. 5.5).

First we suppose that we are interested in finding the trees up to the medium size of the population. We thus compute the top of each tree using the function provided in the lidR package. Then, we simply apply the summary() function and we get about 28 feet of median height of trees.

```
> CrownTop <- find_trees(normlaspt, lmf(ws = 5))
> summary(CrownTop)
```

```
Object of class SpatialPointsDataFrame
Coordinates:
      min        max
x 582360.4    582750
y 5185500.0  5186250
Is projected: NA
proj4string : [NA]
Number of points: 3928
Data attributes:
    treeID          Z                   # note that the
treeID is not a relevant statistics
Min.  : 1.0   Min.  : 2.02
1st Qu.: 982.8   1st Qu.: 15.57
Median: 1964.5   Median: 28.06
Mean  : 1964.5   Mean  : 31.68
3rd Qu.: 2946.2   3rd Qu.: 47.73
Max.  : 3928.0   Max.  : 76.79
```

Then we can map the trees that are up to 28 feets or 8.5 m, and I did so here on top of a raster of the tree heights. Please note the "errors" on the edge as well as the over-estimation with the original dataset.

```
> ttops_28ft <- find_trees(normlaspt, lmf(ws = 28))
```

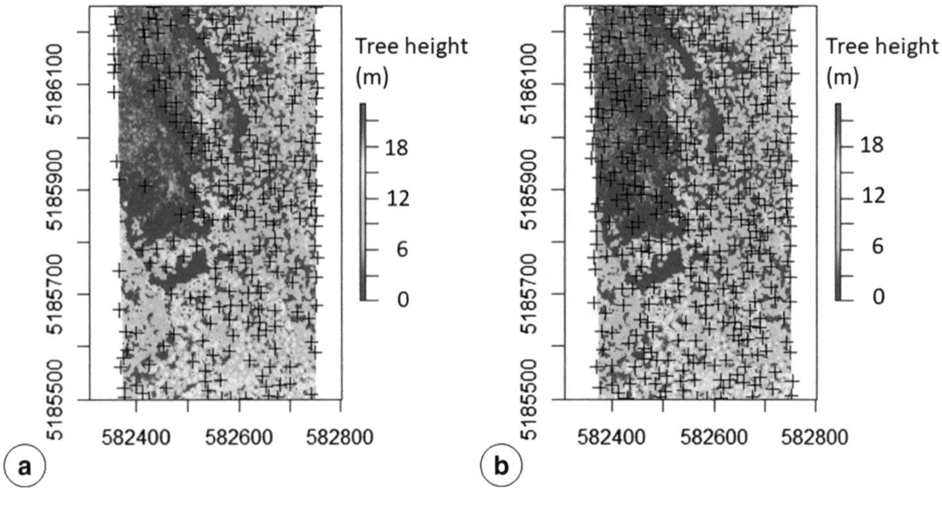

Fig. 5.5 Trees more than 8.5 m in heights calculated from **a** the normalized Lidar dataset and **b** from the original dataset without topographic normalization

```
> ttops_28ftA <- find_trees(normlaspt, lmf(ws = 28))
> par(mfrow=c(1,2))
> plot(chm, col = height.colors(50))
> plot(ttops_28ft, add = TRUE)
> plot(chm, col = height.colors(50))
> plot(ttops_28ftA, add = TRUE)
> plot(nlas)
```

You can also visualize the relation between the topography and the vegetation using a 2D transect by opening the value data of the raster data, and plotting them (Fig. 5.6). In the present example, the rasters I have created from the Lidar data are horizontal lengths of 391 m long, and as I sampled at a 1 pixel per metre rate, a transect only varying in latitude is 391 pixels long (please note that the interpolation can generate errors at the edge, and for my dataset, the first and the last rows are both NAs). Choosing the line that corresponds to a latitude 5,185,900 N, you can reproduce the transect of Fig. 5.6 with the following procedure:

```
# Using this procedure, you can grid the terrain and the
canopy from the LiDAR pointcloud. Please note that
the point-cloud needs to be first classified. If you are
using an "old" LiDAR data that only includes X,Y,Z
or X,Y,Z and the intensity you will need to classify it.
It is true of a point-cloud you could have created from
photogrammetry for instance, where you may not have
intensity values (but you have at least the RGB values
from which you can attempt a classification if it is
```

simple enough, otherwise you may have to burn data from another sensor).

```
> dem <- grid_terrain(las, 1, knnidw())
> dsm <- grid_canopy(las, 1, pitfree(subcircle = 0.2))


> plot(dem@data@values[136850:137241], type=`l',
lwd=2)
> lines(dsm@data@values[136850:137241], col=`-
green', lwd=2)
```

Although this spatial representation provides an intuitive view of the relation between slopes and the vegetation height, it is not a numerical relation as yet. What we want to create is a data plot of the slopes (in Fig. 5.6 it is the topography) with the height of the vegetation (i.e. the green line to which we subtract the black line in Fig. 5.6), and instead of just having a scatter plot, we will use a density plot, showing where certain height trees are concentrated, or in other words on slopes of which order. One will note that to further improve this relation, it would be best to normalize the slope distribution as well. Indeed, if the area you are studying only has slopes < 20°, the chances of finding vegetation on slopes > 20° from your dataset are extremely slim. Similarly, if the slopes > 20° only occupy 2% of your map, you have to reflect it in your final dataset. But for the sake of the demonstration, let's assume this time that it has a homogeneously distributed topography.

In such a case we can represent the relation between slopes (named slp) and the vegetation's height (named

Fig. 5.6 Representation of a cross section along the West–East direction along the 5,185,900 latitudinal coordinate. In black, is the topographic transect generated from the gridded DEM at 1 m resolution, while the green line is the same transect but inclusive of the vegetation

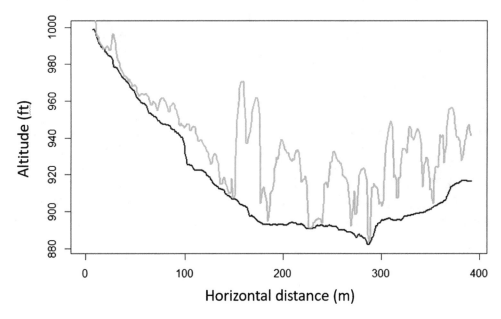

Fig. 5.7 Density plot of the slope (slp) against the height of the vegetation (height) the height is expressed in metres and the slope is in degrees

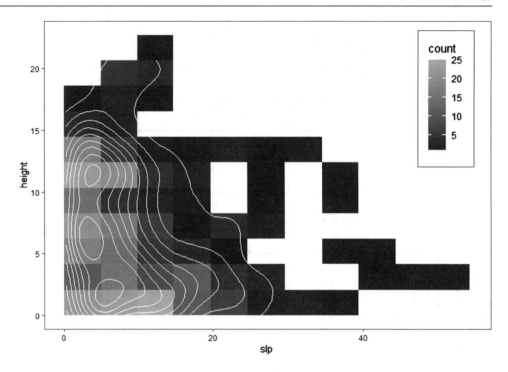

height) using the ggplot library. In the present case, I am loading it with the tidyverse library that has a range of tools for scientific computing (very useful from data manipulation, processing down to graphic generation) (Fig. 5.7):

```r
# For a change, I have written those lines in a R
Notebook, where the code is separated from comments
that you can make in between using ```{r}        ```.
I have chosen to change working locations in R studio,
so that absolute beginners with R and R-studio can get
used to those.


```{r}
library(tidyverse)
a    <-    data.frame(  slp=slopematrix,    height=
(dsmmatrix-demmatrix)*0.3048 )
ggplot(a, aes(x=slp, y=height) ) +
    geom_bin2d(bins = 10) +
    theme_bw() +
    geom_density_2d(colour=`white')
```
```

5.3.5 Micro- and Local Slope Change Example in the Rokko Mountains of Japan

SfM-MVS has without a doubt impacted field mapping and topography more than any other field of geomorphology, but it is also changing outcrop analysis because part of the sediment features or the geologic features can be done back at the laboratory through 3D image analysis, to measure characteristics of fault mirrors from 3D outcrops for instance, or along a fault creating a new way to approach outcrop analysis in the fields of geology and geomorphology.

From the basic observation of outcrops (Fig. 5.8), one can easily design sets of measurement campaigns along forested roads or outcrops on slopes and using the different methods presented in this chapter, follow the evolution of the topographic features, down to the minute scale of a few millimetres.

Conclusion

Point-cloud technology for hillslope monitoring in geomorphology has been and still is a revolutionary tool because acquisition from aircraft and other smaller airborne platforms has allowed the scientists to acquire dense dataset in areas that can be difficult to access, and even monitor (I don't see myself walking a sub-vertical wall with a GNSS cane for instance). Furthermore, slopes are often mantled

Fig. 5.8 Outcrop in the Rokko Mountains above Kobe City near the train station of Suzurandai (Japan)

Sparse point-cloud generation

In Agisoft® Metashape photographs taken from the ground were loaded first and then blur photographs were removed automatically. A set of 10 GCPs with targets that can be automatically recognized by the SfM-MVS software were placed on the outcrop wall. The photographs were taken from feet height, at breast height, and then above my head creating a ~10cm then 1.3 m and then 2.1 m height sets of overlapping photographs, generated using the photographs as explained in the text/

Dense point-cloud generation

This second step uses the MVS algorithm to densify the point-cloud. The dense point-cloud also calculates the color RGB for each point based on the photograph data.

Point-cloud comparison

Leaving the SfM-MVS software (used on the vignette) and using a software like cloud compare, the distance between the generated dense-point and the data acquired using here a total station (the blue flag) is measured. If you are not satisfied with the result, you can constrain the SfM-MVS by adding some of the total station points to the photographs, and then run the first step again, with this time the extra-points.

Meshed- point-cloud

You can then (or not) mesh the point-cloud and then apply the photographs to the surface as a wrap, so that details of the outcrop can be further examined from the laboratory. You can also make measurements of size and orientation of objects from the outcrop.

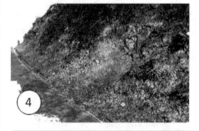

Zoom on the outcrop and one of the target

Example of a detail from the outcrop showing some of the bedrock with joints and fractures, of which the orientation and the dip can be measured for instance. In the detail, you can see the red taped target on the bedrock used to link the total station measurement to the SfM-MVS data (the total station data gives the scale and also the horizontal and the orientation of the outcrop).

with a complex vegetation cover, and when geomorphologists have historically worked in areas with no vegetation because features were then visible, laser point-clouds now provide the geomorphologist with the opportunity to "see" the geomorphology in areas that were otherwise invisible.

Hillslope geomorphology is also gaining new insights, because long-term processes can be measured from features showing important change over long-period of times (for instance the retreat of a rockwall) which divided by a couple of thousands to hundreds of thousand years give us a retreat rate, but we can now combine this measure with high-resolution and accuracy measurement at the decade (and from now on even longer time-spans) measurement of background erosion and event-based erosion, so that we can link long-term change to background erosion (i.e. everyday hardly measurable change) and high-energy singular events.

In other words, HRT point-cloud technology has the potential to solve one of the important questions of geomorphology, namely what process takes the main stage over a long-period of time? Is it the event that we see as extreme on the human-life time-scale, or is it the slow-moving process? In areas where high-energy events are unlikely in present time, it also allows the geomorphologist to ask whether this pattern was different in the past, are we seeing an acceleration or deceleration of erosion. And for this kind of questioning, hillslope is certainly a good geomorphological object to turn to. Finally, this chapter (as for the others) also provided a set of examples that you can do as exercises to train yourself on the "art and craft" of HRT point-cloud technologies, for hillslope of different sizes and from datasets of different origins.

Review Questions

– Why is HRT important to calculate slope stability?

– Is LiDAR calibrated SfM-MVS data matching perfectly the LiDAR data on forested slopes? What are the reasons for potential difficulties and discrepancies?

– Why is vegetation removal from LiDAR data essential for geomorphologic applications? Using a dataset of your own, test the R-code and remove the vegetation.

– How does the vegetation structure correlates with the topography? Explain the importance of LiDAR dataset for this work.

– What are the advantages of SfM-MVS for outcrop measurements?

Problems and Project Ideas

– If you live in the countryside, or near a hilly region where you can find outcrops in sediments, try to use your own camera and a set of targets that you can calibrate using any surveying tool, in order to recreate the 3D of the surface. Measure the size of some features on the outcrops, such as rocks and maybe small tree trunks, and try to calculate them from the pointcloud data. What are the difficulties, what are the limitations? In case you do not have access to surveying equipment, buy a simple tape at the hardware store and anything that will give you a right angle, in such a way that you have a tape in the X direction and one in the Y direction. In the present case, the horizontality does not really matter. You can use those tapes as markers to create the scale of your pointcloud generated by SfM.

– Assuming that you did the first problem or project, plan to go back to the outcrop in a few weeks and then in a few months and re-measure the outcrop surface based on common points that have not moved. Then using CloudCompare, calculate the eroded volume (if any). Do you have errors in your dataset? Reflect on the potential errors.

– If you are a student of geography, geomorphology or geology, you certainly have access to a library of aerial photographs and to some maps. From the overlapping aerial photographs try to recreate a slope with vegetation and one without vegetation. Reflect on the role of the vegetation on the reconstruction of the pointcloud. Think about the potential use of the discrepancies between known topography and the tree canopy height obtained from SfM-MVS on the photographs. If you use historical aerial photographs, can you try to measure the vegetation growth? If so, does it relate to different geomorphological features?

References

1. Ahnert F (1970) Functional relationships between denudation, relief and uplift in large mid-latitude drainage basins. Am J Sci 268:243–263
2. Anderson SW, Anderson SP, Anderson RS (2015) Exhumation by debris flows in the 2013 Colorado front range storm. Geology 43:391–394
3. Anderson RS, Anderson SP (2010) Geomorphology: the mechanics and chemistry of landscapes. Cambridge University Press, 637 p
4. Arnold NS, Rees WG, Devereux BJ, Amable GS (2006) Evaluating the potential of high-resolution airborne LiDAR data in glaciology. Int J Remote Sens 27:1233–1251
5. Bartley R, Bainbridge ZT, Lewis SE, Kroon FJ, Wilkinson SN, Brodie JE, Silburn DM (2014) Relating sediment impacts on coral reefs to watershed sources, processes and management: a review. Sci Total Environ 468–469: 1138–1153
6. Barton NR, Choubey V (1977) The shear strength of rock joints in theory and practice. Rock Mech 10:1–54
7. Benn DI, Evans DJA (2010) Glaciers and glaciation, 2nd edn. Routledge Taylor and Francis, 816 p
8. Bennett MR, Glasser NF (2009) Glacial geology—ice sheets and landforms, 2nd edn. Wiley-Blackwell, 385 p
9. Burns WJ, Coe JE, Kaya BS, Ma L (2010) Analysis of elevation changes detected from multi-temporal LiDAR surveys in forested landslide terrain in western Oregon. Environ Eng Geosci 16:315–341
10. Corsini A, Borgatti L, Cervi F, Dahne A, Ronchetti F, Sterzai P (2006) Estimating mass-wasting processes in active earth slides—earth flows with time-series of high-resolution DEMs from photogrammetry and airborne LiDAR. Nat Hazard 9:433–439
11. Crozier MJ, Gage M, Pettinga JR, Selby MJ, Wasson RJ (1992) The stability of hillslopes, Chap 3. In: Soons MJ, Selby MJ (eds) Landforms of New Zealand, pp 45–66
12. Culling WEH (1963) Soil creep and the development of hillslopes. J Geol 71:127–161
13. De Blasio FV (2011) Introduction to the physics of landslides. In: Lecture notes on the dynamics of mass wasting. Springer, 408 p
14. Domenech G, Corominas J, Mavrouli O, Merchel S, Abellan A, Pavetich S, Rugel G (2018) Calculation of the rockfall recession rate of a limestone cliff, affected by rockfalls, using cosmogenic chlorine-36. Case study of the Montsec Range (Eastern Pyrenees, Spain). Geomorphology 306:325–335

15. Eustace AH, Pringle MJ, Denham RJ (2011) A risk map for gully locations in central Queensland, Australia. Eur J Soil Sci 62–3:431–441

16. Evans IS (1972) Cartographic automation and geographic analysis. Area 4:64-66

17. Foroutan M, Marshall SJ, Menounos B (2019) Automatic mapping and geomorphometry extraction technique for crevasses in geodetic mass-balance calculations at Haig Glacier, Canadian Rockies. J Glaciol 65(254):971–982

18. Gallen SF, Wegmann KW, Frankel KL, Hughes S, Lewis RQ, Lyons N, Paris P, Ross K, Bauer JB, Witt AC (2011) Hillslope response to knickpoint migration in the Southern Appalachians: implications for the evolution of post-orogenic landscapes. Earth Surf Process Landforms 36:1254–1267

19. Gomez C, Hotta N (2021) Deposits' morphology of the 2018 Hokkaido Iburi-Tobu earthquake mass movements from LiDAR and aerial photographs. Remote Sens 13:3241. https://doi.org/10.3390/rs13173421

20. Gomez-Gutierrez A, Schnabel S, Brenguer-Sempere F, Lavado-Contador F, Rubio-Delgado J (2014) Using 3D photo-reconstruction methods to estimate gully headcut erosion. CATENA 120:91–101

21. Goodwin NR, Armston J, Stiller I, Muir J (2016) Assessing the repeatability of terrestrial laser scanning for monitoring gully topography: a case study from Aratula, Queensland, Australia. Geomorphology 262:24–36

22. Hack JT (1973) Stream-profile analysis and stream-gradient index. J Res U.S. Geol Surv 421–429

23. Hartmeyer I, Keuschnig M, Delleske R, Krautblatter M, Lang A, Schrott L, Prasicek G, Otto J-C (2020) A 6-year lidar survey reveals enhanced rockwall retreat and modified rockfall magnitudes/frequencies in deglaciating cirques. Earth Surf Dyn 8:753–768

24. Harvey AM, Mather AE, Stokes M (2005) Alluvial fans—geomorphology, sedimentology, dynamics. Geol Soc Spec Publ 251:248 p

25. Hayakawa YS, Oguchi T (2009) GIS analysis of fluvial knickzone distribution in Japanese mountain watersheds. Geomorphology 11:27–37

26. Hayakawa Y, Matsukura Y (2003) Recession rates of waterfalls in Boso Peninsula, Japan and a predictive equation. Earth Surf Process Landforms Short Commun 28:675–684

27. Hungr O, Fell R, Couture R, Eberhardt E (2005) Landslide risk management. Abema-CRC Press, 764 p

28. Imaizumi F, Masui T, Yokota Y, Tsunetaka H, Hayakawa YS, Hotta N (2019) Initiation and runout characteristics of debris flow surges in Ohya landslide scar, Japan. Geomorphology 339:58–69

29. Imaizumi F, Nishii R, Murakami W, Daimaru H (2015) Parallel retreat of rock slopes underlain by alternation of strata. Geomorphology 238:27–36

30. Imaizumi F, Tsuchiya S, Ohsaka O (2005) Behaviour of debris flows located in a mountainous torrent of the Ohya landslide, Japan. Can Geotech J 42:919–931

31. Jones RR, Kokkalas S, McCaffrey KJW (2009) Quantitative analysis and visualization of nonplanar fault surfaces using terrestrial laser scanning (LIDAR)—the Arkitsa fault, central Greece, as a case study. Geosphere 5:465–482

32. Karamitros I, Ganas A, Chatzipetros A, Valkaniotis S (2020) Non-planirity, scale-dependent roughness and kinematic properties of the Pidima active normal fault scarp (Messinia, Greece) using high-resolution terrestrial LiDAR data. J Struct Geol 136:104065, 1–15

33. Kohv M, Talviste P, Hang T, Kalm V, Rosentau A (2014) Slope stability and landslides in proglacial varved clays of western Estonia. Geomorphology 106:315–323

34. Kokkalas S, Jones RR, McCaffrey KJW, Clegg P (2007) Quantitative fault analysis at Arkitsa, Central Greece, using terrestrial laser-scanning (LiDAR). Bull Geol Soc Greece 37:1–14

35. Lague D, Brodu N, Leroux J (2013) Accurate 3D comparison of complex topography with terrestrial laser scanner: application to the Rangitikei canyon (N-Z). ISPRS J Photogramm Remote Sens 82. https://doi.org/10.1016/J.ISPRSJPRS.2013.04.009

36. Marin RJ, Velasquez MF (2020) Influence of hydraulic properties on physically modelling slope stability and the definition of rainfall thresholds for shallow landslides. Geomorphology 351:106976, 1–14

37. Mason J, Schneiderwind S, Pallikarakis A, Wiatr T, Mechernich S, Papanikolaou I, Recherter K (2016) Fault structure and deformation rates at the Lastros-Skafa Graben, Crete. Tectonophysics 683:216–232

38. Mayr A, Rutzinger M, Bremer M, Oude Elberink S, Stumpf F, Geitner C (2017) Object-based classification of terrestrial laser scanning point cloud for landslide monitoring. Photogram Rec 32:377–397

39. McInnes R, Jakeways J, Fairbank H, Matie E (2007) Landslides and climate change—challenges and solutions. In: Proceedings of the international conference on landslides and climate change, 512 p

40. Mergili M, Marchesini I, Rossi M, Guzzetti F, Fellin W (2014) Spatially distributed three-dimensional slope stability modelling in a raster GIS. Geomorphology 206:178–195

41. Middleton M, Nevalainen P, Hyvonen E, Heikkonen J, Sutinen R (2020) Pattern recognition of LiDAR data and sediment anisotropy advocate a polygenetic subglacial mass-flow origin for the Kemijarvi hummocky moraine field in northern Finland. Geomorphology 362:107212, 1–17

42. Miller H, Cotterill CJ, Bradwell T (2014) Glacial and paraglacila history of the Troutbeck Valley, Cumbria, UK: integrating airborne LiDAR, multibeam bathymetry, and geological field mapping. Proc Geol Assoc 125:31–40

43. Montgomery DR, Dietrich WE (1988) Where do channels begin? Nature 336:232–234

44. Morgenroth J, Gomez C (2014) Assessment of tree structure using a 3D image analysis technique—a proof of concept. Urban For Urban Green 13:198–203

45. Nolan M, Larsen C, Sturm M (2015) Mapping snow depth from manned aircraft on landscape scales at centimeter resolution using structure-from-motion photogrammetry. Cryosphere 9:1445–1463

46. Ojala AEK (2016) Appearance of De Geer moraines in southern and western Finland—implications for reconstructing glacier retreat dynamics. Geomorphology 255:16–25

47. Panek T, Minar J, Vitovic L, Brezny M (2020) Post-LGM faulting in Central Europe: LiDAR detection of the >50 km-long Sub-Tatra fault, Western Carpathians. Geomorphology 364:107248, 1–15

48. Patton FD (1966) Multiple modes of shear failure in rock and related materials. PhD thesis, University of Illinois Urbana-Champaign, published in 1972, 564 p

49. Pelto BM, Menounos B, Marshall SJ (2019) Mutli-year evaluation of airborne geodetic surveys to estimate seasonal mass balance, Columbia and Rocky Mountains, Canada. Cryosphere 13:1709–1727

50. Pierson LA, Davis SA Van Dickle R (1990) Rockfall hazard rating system implementation manual. Federal Highway Administration. FWHA-OR-EG-90-01

51. Pierson LA, Van Vickle R (1993) Rockfall hazard rating system: participants' manual. Federal Highway Administration. FHWA-SA-93-057

52. Plackowska E, Cebulski J, Bryndza M, Mostowik K, Murawska M, Rzonca B, Siwek J (2021) Morphometric analysis of the channel heads based on different LiDAR resolutions. Geomorphology 375:107546, 1–10

53. Purdie H, Rack W, Anderson B, Kerr T, Chinn T, Owens I, Linton M (2015) The impact of extreme summer melt on net accumulation of an avalanche fed glacier, as determined by ground-penetrating radar. Geogr Ann 97(4):779-791. https://doi.org/10.111/geoa.12117

54. Purdie H, Bealing P, Gomez C, Anderson B, Marsh OJ (2020) Morphological changes to the terminus of a maritime glacier during advance and retreat phases: Fox Glacier/Te Moeka o Tuawe, New Zealand. Geogr Ann Ser A. https://doi.org/10.1080/04353676.2020.1840179

55. Roberts GP, Ganas A (2000) Fault-slip directions in central and southern Greece measured from striated and corrugated fault planes: comparison with focal mechanism and geodetic data. J Geophys Res Solid Earth 105:23443–23462

56. Różycka M, Jancewicz K, Migon P, Szymanowski M (2021) Tectonic versus rock-controlled mountain fronts—geomorphometric and geostatistical approach (Sowie Mts., Central Europe). Geomorphology 373:107485, 1–20

57. Scaioni M, Longoni L, Melillo V, Panini M (2014) Remote sensing for landslide investigations: an overview of recent achievements and perspectives. Remote Sens 6:9600–9652

58. Schmidt J, Dikau R (2004) Modeling historical climate variability and slope stability. Geomorphology 60:433–447

59. Sookhan S, Eyles N, Putkinen N (2016) LiDAR-based volume assessment of the origin of the Wadena drumlin field, Minnesota, USA. Sed Geol 338:72–83

60. Terzaghi KV (1943) Theoretical soil mechanics. Wiley, 510 p

61. Tindall D, Marchand B, Gilad U, Goodwin N, Denham R, Byer S (2014) Gully mapping and drivers in the grazing lands of the Burdekin catchment. RP66G Synthesis Report, Queensland Department of Science, Information Technology, Innovation and the Arts. Brisbane, Australia

62. Trommelen MS, Ross M, Ismail A (2014) Ribbed moraines in northern Manitoba, Canada: characteristics and preservation as part of a subglacial bed mosaic near the core regions of ice sheets. Quat Sci Rev 87:135–155

63. Tseng C-M, Lin C-W, Stark CP, Liu J-K, Fei L-Y, Hsieh Y-C (2013) Application of a multi-temporal, LiDAR-derived, digital terrain model in a landslide-volume estimation. Earth Surf Process Landforms 38:1587–1601

64. Tsuchiya S, Imaizumi F (2010) Large sediment movement caused by the catastrophic Ohya-kuzure landslide. J Disaster Res 3:257–263

65. Wasklewicz TA, Hattanji T (2009) High-resolution analysis of debris flow-induced channel changes in a headwater stream, Ashio Mountains, Japan. Prof Geogr 61:234–249

66. Wawrzyniec TF, McFadden LD, Ellwein A, Meyer G, Scuderi L, McAuliffe J, Fawcett P (2007) Chronotopographic analysis directly from point-cloud data: a method for detecting small, seasonal hillslope change, Black Mesa Escarpment, NE Arizona. Geosphere 3(6):550–567

67. Wiatr T, Reicherter K, Papanikolaou I, Fernandez-Steeger T, Mason J (2013) Slip vector analysis with high resolution t-LiDAR scanning. Tectonophysics 608:947–957

68. Wilkinson M, McCaffrey KJW, Roberts G, Cowie PA, Phillips RJ, Michetti AM, Vittori E, Guerrieri L, Blumetti AM, Bubeck A, Yates A, Sileo G (2010) Partitioned postseismic deformation associated with the 2009 Mw 6.3 L'Aquila earthquake surface rupture measured using a terrestrial laser scanner. Geophys Res Lett 37:L10309, 1–7

69. Yamada S (1999) The role of soil creep and slope failure in the landscape evolution of a head water basin: field measurements in a zero order basin of northern Japan. Geomorphology 28:329–344

Pointcloud and Volcanic Geomorphology

6

Abstract

This chapter turns towards a specific type of hillslopes: volcanoes and how pointclouds can be used in volcanic geomorphology. The chapter first presents general features of volcanic geomorphology, before presenting the use of SfM-MVS and LiDAR to deal with the specific difficulties of working in volcanic environment (i.e. recording steep craters' slopes or staying away from an erupting vent …). This chapter also uses this opportunity to present how pointcloud technologies can be used for micro-geomorphology (in the present case, from the example of a pyroclastic flow deposit surface differential erosion) as well as micro-geomorphology in the laboratory (in the example given, a small sand volcanoes). From the examples, the chapter also proposes a short discussion of the role of grid cell size and sampling strategy from a volcanic geomorphology dataset processed in R.

Learning Outcomes: Overcoming the Difficulties of Working in "Erupting" Environments

After reading this chapter, you will understand the relationship between different volcanic landform types and the right acquisition tools to obtain data and information to be processed into a pointcloud. You will also be able to identify the difficulties inherent in volcanic environments (i.e. access difficulties, low reflectivity of the material (or not), sub-vertical features, e.g. caldera walls). A set of worked examples also show you how you can bring your UAV to the field and in combination with other instruments how you can then verify the quality of your work (even with some very cheap solutions, if you are only interested in relative localization).

Content and Objectives

This chapter focuses on the specific aspects of working with and creating pointclouds to depict the geomorphology of volcanoes. The chapter is articulated around a quick introduction to some key volcanic geomorphological features from the pointcloud perspective (it is thus not an exhaustive list), before progressively zooming on different objects, such as the crater for instance, before ending with micro-geomorphology. Following a similar flavour given to the other chapters, the presentation of the different topics is supported by literature examples, a few worked examples and a list of extra-readings that you can go and consult to study further.

The objectives of this chapter are to:

(1) Give an understanding of the specific difficulties of working on volcanoes and what are the best adapted tools and methods.
(2) Let the use know how to work and know what can be done with pointcloud on volcanoes.
(3) Let the student learn about some of the main features that can be easily measured for a short research project, so that she/he might develop interest in working with volcanic material.

Who is the chapter for? If you are a student of volcanic geomorphology, or a volcanologist who is discovering the potentials of pointcloud technologies for your field, this is certainly the chapter for you. If you are a specialist of volcanology, you may want to skip the first section that describes some generalities about volcanoes and concentrate on the different examples as well as the last section that presents a worked example. In this chapter, as there isn't any novel tool usage, I invite the reader to go to Chaps. 2 and 3 to read about the nitty–gritty details of data processing, error analysis, etc.

Who is the chapter not for? If you are a veteran of volcanic geomorphology and you have done a lot of high-resolution topography data acquisition and processing, maybe this chapter will not teach you anything that you already know.

© Springer Nature Switzerland AG 2022
C. Gomez, *Point Cloud Technologies for Geomorphologists*,
Springer Textbooks in Earth Sciences, Geography and Environment.
https://doi.org/10.1007/978-3-031-10975-1_6

Introduction

Bolt et al. wrote in their classic book on Geological hazards in 1975 that "Volcanic eruptions are among the most violent, spectacular, and awe-inspiring manifestations of nature. It is hardly surprising, therefore, that from time immemorial they have fascinated man, invoking in him sometimes terror, sometimes religious veneration, sometimes artistic appreciation of beauty, and always respect." (p. 63). This fascination has consequently attracted generations of scientists and amateurs to measure volcanoes and notably their morphology, like the postal worker in Hokkaido, Japan, who, in the middle of Second World War, took on himself to record the growth of Mt. Showa Shinzan, which emerged from a barley field between 1943 and 1945. The set of profile changes penned on a sheet of paper from a window of the post office provides an invaluable insight into the growth of the volcano, but also in the interest and fascination volcanoes exert on people. For us, interested in morphology and topography, and notably the topographical changes triggered by volcanic activity, it also shows the central place morphological change plays. In the "volcanoes" section of his review, Tarolli [41] shortly presents the different laser-based tools (ALS, TLS) linking different measurements to the understanding of the processes the measured deposits explained. He also states that "high-resolution topographies can improve the understanding in hazard and risk assessment". High-resolution topography of volcanoes is therefore a tool for the scientists and also for the practitioners to make informed decisions.

6.1 A Short Introduction to the Geomorphology of Volcanoes

6.1.1 Birth and Differentiation of Volcanic Structures

Volcanic rocks, pyroclastic rocks and plutonic rocks occupy respectively 6%, 0.6% and 7% of the Earth surface [25], which is the domain—not solely—where geomorphologists work. For a volcano, this slice of the Earth is only the tip of the iceberg, but, it is a tip from which hazards and disaster risk models are often based on, and where the majority of the evidences for the volcano evolution are drawn from—if not for a few very rare deep drilling projects, like at Unzen Volcano in Japan [39], or in the grabben of Merapi Volcano in Indonesia [14]. Otherwise, all the data have to be remotely sensed through various geophysical methods, until scientists invent a method to go and retrieve samples in "deep overheated, over-pressured" environments. Volcanic geomorphology is thus an extremely precious dataset as it is one of

the few pieces of information that is really accessible, and it is important to extract all the potential information from it.

Whether you directly study geomorphology, use the data in a pyroclastic or lava or lahar flow model, or you are trying to understand the sedimentary formations, geomorphology and HRT geomorphology that are essential components to understand volcanic structures and environments in space and time.

6.1.2 Erosion, Dismantlement and Change in Volcanic Structures

The approach to volcanic erosion and dismantlement can be grossly divided between (a) a local-measures and process-based approach (how much erosion and deposition, and sediment transport occurred with one or a series of lahar events for instance [28]), and (b) a longer-term approach, which is presently lead by the work of Professor Lahitte who is developing specialized-tools for this purpose (e.g. [27]), and which combines surface dating and models of morphological evolution and high-resolution topography, which is part of a long-tradition—yet very active—of palaeo-surface reconstruction in geomorphology (e.g. [24]).

(a) "Short-term" erosion, sedimentation and landform construction processes. Short-term processes are best studied on structures that change rapidly, like in the aftermath of an explosive eruption where large amounts of fragmented pyroclasts are rapidly remobilized, especially under humid climates. Watersheds dominated by volcanic activity can produce sediment yields in the order of 10^4–10^7 t/km^2 [30, 40]. In detail, the related erosion/deposition cycles are, however, rather irregular, notably due to the change in the available sediment stock and eventually the rainfall patterns [42].

(b) "Long-term" processes. Longer-term measure of evolution of volcanic structure with geomorphology looks into processes at the ten of thousand years to the million years scale. The methodology is different from the "short-term" processes mentioned above as there are no tools or human eyes that can follow a million years of change. Therefore, the method relies more on interpolation, models and processes that are at a short-term scale, of very little importance. In early computational work on the topic, Leverington et al. [29] used GIS to compensate for deformation of the "isostatic water plane" and then recreate the palaeo-topography. Thanks to the development of topographic dataset such as the SRTM, regional approaches for calculation on dozens of structures concomitantly have emerged. Karatson et al. [24] have proposed erosion volumes and rates for the

Fig. 6.1 DEMs of **a** the Aso Caldera (South Japan) and **b** the Izu-Oshima Caldera (South of Tokyo City and part of Tokyo prefecture). Both caldera display complex forms of the collapsed-floor caldera, with a flat bottom, a steep-sloped surrounding rim and numerous complexities. The Aso Caldera is 15 km east–west and 20 km north–south, while the Izu-Oshima is 15 km from tip to tip and the caldera is about 2.5 km diameter

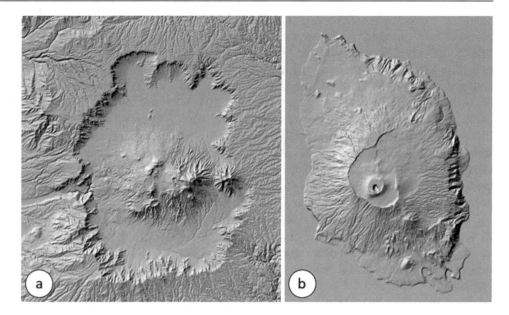

Neogene to Quaternary stratovolcanoes in the Andes Cordillera. At a smaller scale, zooming on one to a small group of structures, Lahitte et al. [27] reconstructed the paleo-surfaces of volcanic structures of the southern part of Basse-Terre in the Guadeloupe archipelago, with a timeline starting as early as 650,000 years ago. Dibacto et al. [8] used a similar approach to 4 volcanic groups divided in 17 volcanic sub-structures located between the Transylvanian basin and the East Carpathian fold and thrust range with dating ranging from 1.9 to 6.7 million years.

6.1.3   Morphologies and Landforms of Volcanoes

The general morphology of volcanic structures firstly depends on how high the magma rises to the Earth surface, and on its viscosity, which is itself controlled by the chemical composition—broadly separating hotspot oceanic volcanism, rift volcanism and plate margin volcanoes. For a geomorphologist, this results in a division of the structures' shapes and size and which in detail are modified depending on the surrounding environment (geologic, topographic and climatic). Karlstrom et al. [25] classified volcanic environments based on the geomorphology, by separating landforms generated by intrusions, from caldera subsidence, from those generated by volcanic eruptions (effusive and explosive). Furthermore, volcanoes seldom show a "textbook" landform, they are usually a combination of effusive and explosive eruption deposits and formation. For instance, calderas are more than just a rim surrounding a collapsed floor, they display irregular and complex morphologies like

at Aso Volcano in South Japan and at Izu-Oshima to the South of Tokyo (Fig. 6.1).

Although the description of landforms in geomorphology started with hand sketches and qualitative descriptions, the development of digital elevation models and point-cloud technologies has compelled geomorphologists to develop numerical methods to characterize landforms [18] and to separate features at different scales using Fourier transforms and wavelets [13].

6.1.4   Specificities of Pointcloud Data for Volcanic Structures

Volcanoes, and more especially active volcanoes, are most certainly a target of choice for topographic data acquisition using remote sensing methods such as laser technologies and photogrammetry, because they both allow for the acquisition of data from a remote and safe position (even terrestrial laser scanners can have range of several kilometres). For active volcanoes, ALS mounted on a UAV is the tool of choice, because it allows for the operator to be several kilometres away, without having anyone flying in a plane and thus being in harm's way, especially with explosive eruptions. UAV mounted with cameras is also a possibility, but the gas escaping the vent and the smoke often provide poor visibility, in such a way that it is certainly possible to image the gas plume [15], but it is more difficult to image what is underneath.

Furthermore, erupting volcanoes are modifying their geometries, in such a way that even with an array of high-resolution GNSS and other single-point measuring devices, a broad range of the geometric change has to be ignored. Pointcloud technologies are now allowing for the scientists to follow in detail the evolution from one point-cloud to another, providing for instance that the pointclouds are tied each time to high-spatio-temporal-resolution benchmarks, such as GNSS measured targets. At Telica Volcano, in the Maribios Range of Nicaragua, SfM has been used from flights taken between 2011 and 2017, in order to measure changes in the crater. The authors used between 13.7 and 60.5 million points' pointclouds and compared each of them with the M3C2 algorithm in CloudCompare. The results have shown that the crater wall have retreated by 15 to more than 40 m in a year span, and that >35 m increase in the elevation of the crater floor was due to assumed magma uprise. In 2011, a ∼50 m floor-rise, has been attributed to debris deposits on the crater floor [19]. In this typical example of an active explosive volcano, crater evolution and mapping is therefore best done by UAV and UAV photography for SfM-MVS purposes (providing that there is enough visibility). One advantage of using a small drone with a camera over an ALS drone is the cost of the equipment if you lose it due to turbulence or some incidents over the vent.

6.2 Selected Landforms and Adapted Point-Cloud Technologies

Like for the other chapters, the set of landforms selected in this section is not an exhaustive list—which may be the topic of a book on volcanoes (although encyclopaedic knowledge has left the place to more mechanistic approaches that are not lists of objects anymore). It is not a comprehensive description of volcanic landforms either, but rather a state of the art on how scientists are using point-cloud in their research in volcanic geomorphology. Because of this particular focus, I have divided the volcanic landforms from the volcanic scale down to the local feature scale, dividing surfaces based on how close to the NADIR direction they are. This choice was motivated by the fact that both scales and verticality of slopes will have a strong influence on the tools and methods of data construction and representation.

6.2.1 Volcano-Scale Geomorphology and Point-Cloud Technologies

Difficulties	– Large-scale features spanning for several tens to hundred square kilometres – Potential cover of vegetation and smoke/dust during and just after eruptions
Recommended tools (not exhaustive)	– ALS from aeroplane or from satellite station – Aircraft-based aerial photographs

As physical models of flows occurring from and on volcanoes (e.g. lahar or pyroclastic flow models) are increasingly present in process geomorphology, and/or for natural hazards and disaster risk management, high-resolution DTMs and DEMs are essentials to provide those models with the most accurate boundary conditions possible. Furthermore, topographic data also provides temporal snapshots of active volcanoes, so that their evolution, even during an eruption event can be measured and followed [38].

Thanks to airborne data acquisition systems (mostly conventional aircraft LiDAR), volcanic structures in remote locations can also be studied with high precision and accuracy, as it has been done at Erebus Volcano in Antarctica, where the NASA collected ALS data, providing a unique topographic dataset of the active crater of the volcano [4]. This research also confirmed the presence of dykes and faults that were previously mapped at a lower resolution, in such a way that the dataset is a baseline to survey the morphological change of the volcano. Thinking over a 50–100 years' period, point-cloud technologies are now providing dataset at such a resolution that future scientists will be able to use those dataset as baselines for long-term monitoring. Up to present, historical data was either GPS/GNSS data or topographic maps. Even high-precision and high-accuracy single topographic control points are relatively recent advances thanks to GNSS and differential GNSS data acquisition. Therefore, they are often available over only a couple of decades. And data generated 50–100 years ago is mostly topographic maps that are most often not sufficient to quantify change during that span. Completing the progress of laser technology, photogrammetry can also be used to reconstruct the volcanic structures. Chronological evolution from aerial photography is most certainly a possibility, but the accuracy is seldom below a couple of metres, due to (1) a lack of control points in ever-changing landscapes and (2) photographs that were not taken for the purpose of digital photogrammetry [16].

At Etna Volcano, the LiDAR recorded in 2005 [9] provides a dataset covering the entire volcano, with ∼257 million points over an area of 612 km². It has a calculated accuracy inferior to 40 cm vertically [2, 9]. Because of the point density and the high accuracy, the data could then be further mined to investigate sub-features, such as the scoria cones that are on the volcanic structure [10]. The size of this dataset does not, however, belong to a large LiDAR dataset. Volcano Mazama surrounds Crater Lake national parks in Oregon (USA) and the 10 km diameter caldera, and the volcanic structure was documented with a >5 billion points LiDAR dataset [36].

A LiDAR data is however not solely a huge set of X, Y, Z coordinates, and it is also a record of other parameters, such as the intensity of the laser return for instance. This parameter can be used to differentiate the type of surface formations that are being detected—besides the local morphology and surface roughness of the point-cloud [12]. Are we detecting a lava deposit topography or a volcaniclastic granular deposit, etc. are questions we can also ask the LiDAR data. This capacity can also be further enhanced by combining LiDAR data with hyperspectral imaging, so that you can add to your topographic data, information about the mineral distribution and the lithological characteristics [26].

From the dataset, different types of metrics can then be extracted for investigating the volcanic structure, using for instance derived hypsometric lines on scoria cones [47] and on stratovolcanoes notably [13] using tools that come from the field of signal processing.

6.2.2 Slopes and Sub-vertical Features (Flanks and Crater Walls)

Difficulties	– Sub-vertical features difficult to capture from NADIR sensors – Sometimes, local features like craters only a few tens to hundred metres in diameter – Stratovolcanoes' valleys are often deeply incised but narrow, limiting the field of view on the walls
Recommended tools (not exhaustive)	– TLS – Hand-held cameras photography for photogrammetry – Small UAV photography with tilted camera

Point-clouds and DEMs data created at the scale of a whole volcano is therefore best recorded using fly-over techniques, especially when the volcano is active or erupting. But, as it is the case for cliff faces in coastal geomorphology, aerial data

acquisition is not always sufficient to capture the details of sub-vertical features, because the angle of the laser array or the direction of the vectors between the camera sensor and the investigated surface is almost parallel to one another. For instance, the sub-vertical surfaces are characteristics of lahar valleys, which can be incised several tens of metres deep for only a couple of metres' width. Another typical example of sub-vertical features on a volcano is the volcanic vents, craters and caldera walls notably (Fig. 6.2).

Although volcanoes encounter dramatic topographic changes during eruptions, secondary processes that can be related to what occurs on non-volcanic slopes occur, and furthermore topographic changes linked to pre-eruptive seismicity and other "secondary processes" also occur. At Mt. Vesuvius, Pesci et al. [34] have observed rockfall movements starting from the crater walls for the years 2005, 2006 and 2009 using a terrestrial laser scanner (Optech ILRIS 3D, with a 1535 nm, with a pulse frequency of 2.5 kHz and a beam divergence of 0.17 mrad). The data was georeferenced using a digital DTM obtained from photogrammetry, and the point-clouds were superimposed to one another using a locked georeferenced frame. From this dataset, the authors generated deformation maps of the caldera walls, so that portions of the sub-vertical crater wall where material moved from and the surface of the apron underneath where material accumulated could be detected ([34]: Fig. 4). The author calculated that the error due to misalignment between the different scans was eventually in the range of 3–4 cm, but that the greatest error was introduced by the triangulation process, which simplified the model to 30 cm face-length triangles. If we were looking at a built structure (like a bridge), a 30 cm error would be inacceptable, but for most of the geomorphological purposes, it still fits within the realm of a high resolution and accuracy dataset—I often remind my students that only 20 years ago, any topographic data with an accuracy and precision <1 m

Fig. 6.2 Main vent of Mt. Mihara in the caldera of Izu-Oshima (cf. Fig. 6.1b). The data was collected using UAV-based photogrammetry (SfM-MVS) in 2017. The left panel shows the vent with the altitude with a colour scale, while the right side of the figure shows the same data but in RGB colours. One can see the results of wall collapses and the apron at the toe of the sub-vertical slopes

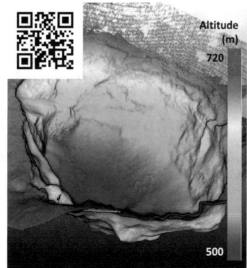
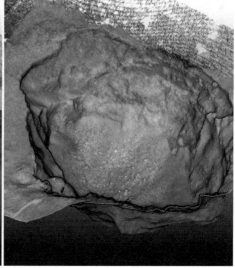

would have been a great achievement. This work follows on earlier work, in which the authors combined aerial photographs, photogrammetry and TLS to create a model at 0.3–0.5 m resolution, when the all Vesuvius Volcano photogrammetric model had a resolution of 2.5 m [33]. In a more remote setting at Erebus Volcano in Antarctica, a combination of ALS and TLS surveys performed respectively in 2001 and in 2010 also showed that the crater wall had moved out by more than 20 m on the SW transect, while the floor with the lava lake and the Werner Vent moved down by about 25 m [23]. They even completed this longer-term assessment with smaller-scale changes showing a cyclic pattern of relative lake-level change that oscillated between ~ 1 and ~ 3 m vertically, with periods in the order of 15 min, giving an insight into the driving "pulses" in the magma column (Fig. 7 in Jones et al. [23]). At Merapi Volcano in Indonesia, similar work on the evolution of the crater and the dome inside was conducted using UAV-based photogrammetry between 2012 and 2016 using a small commercial quadcopter to perform SfM-MVS photogrammetry [5].

Finally, volcanic slopes are also under the influence of processes that one encounter in mountain areas, and the dominance of one type of process over another is controlled by the period of return between each eruption and the erosion processes acting on the quiescent volcano.

6.2.3   Lava Domes, Pyroclastic-, Lava Flows, Lahar Deposits

Difficulties	– Highly reflective surfaces (andesites) or very homogenous black surfaces not conductive to automation in photogrammetry
	– Features active and rapidly eroded, requiring acquisition when the volcano is often still at risk to erupt further
	– The need to collect GCPs in hazardous areas
Recommended tools (not exhaustive)	– TLS
	– UAV and helicopter-based LiDAR
	– Small UAV photography for photogrammetry with extensive markers

Although collecting together pyroclastic lava flow and lahar deposits, and lava domes may just seem an odd series for any volcanologist and geologist, we are here following a topographic and morphologic logic, with units that are spatially conspicuous, and which can sometimes be fully or partially trapped in a low topography (for instance, inside a crater for the volcanic dome, or inside a valley for the pyroclastic deposits or the lava deposits).

Lava domes—Lava domes are the surface manifestation of a viscous magma that reaches the surface in the shape of a

semi-hemisphere. The dome can crack under the growth of others underneath, in which case the material can travel down the volcanic slopes. Domes can be found on andesitic and rhyolitic volcanoes notably, because of the viscous lava they produce. Although some domes, like at the Unzen Volcano, Japan (Fig. 6.3), tend to collapse under the effect of gravity, others explode under the increased gas pressure, suddenly releasing high-pressure magma leading to the expulsion of fragmented material. Measuring their growth and evolution is therefore essential for both scientists and hazards and disaster risk managers.

Using DEMS pre- and post-eruption (2014 and 2017) De Beni measured the topographic change created by a lava flow at Mt. Etna. They calculated a volume of 1.4×10^6 m^3 $\pm$ 20% of emplaced lava deposits [6]. This work was conducted using UAV-based SfM-MVS photogrammetry and a GNSS (Leica Zeno 10) for terrestrial survey. Because pointcloud technologies have a high-density footprint, they are also an excellent tool for measuring the changes and movement of discrete features, such as large blocks. At Unzen Volcano (Fig. 6.3), the sliding dome is surmounted by several metres high blocks that tower among the other debris, and as the dome is sliding and the surface is being eroded and is "expanding", the large blocks also move (Fig. 6.4—cf. the yellow circles) and eventually break as it occurred in 2021.

Confined and semi-confined deposits: lahars and ignimbrites—Ignimbrites are the deposits from pyroclastic flows (or more commonly named pyroclastic density currents in recent literature), and they can be either confined in valleys or spreading from the valleys and outside, or simply over open slopes. These deposits are well described in Branney and Kokelaar [3]. Often starting by the remobilization of ignimbrites (but not only), lahars are mixtures of water and debris (mostly clastic but also including bio-debris) that flow on or from the slopes of a volcano and that have the particularity of carrying heterometric materials in a dense matrix. There are plenty of classifications based on the deposit characteristics, the flow mechanics and rheology, the sediment concentration, etc., but we can all accord on the fact that they leave in the talweg and sometimes around them deposits that can be several tens of metre thick and that can ensue from a highly erosive flow, deeply modifying the talwegs that radiates from volcanoes. At Colima Volcano (Mexico), Hurricane Patricia in 2015 brought an estimated 500 mm of accumulated rainfall over 24 h, triggering lahars that eroded as much as 5–6 m vertically [44]. Ignimbrites, whether they are valley confined or the results of flows on "open slopes" typically display a lobe at their downstream end, with various details such as lateral levees, which from LiDAR data has been used to infer deposition velocities. For the 1993 Lascar pyroclastic flow

Fig. 6.3 Volcanic dome of
Unzen Volcano on Shimabara
Peninsula (Japan) acquired
through the monitoring
programme funded by the
Japanese government, in the
aftermath of the last eruption and
as part of a research programme
the author had joined, and which
was directed by Miyazaki
University and Dr Shinohara

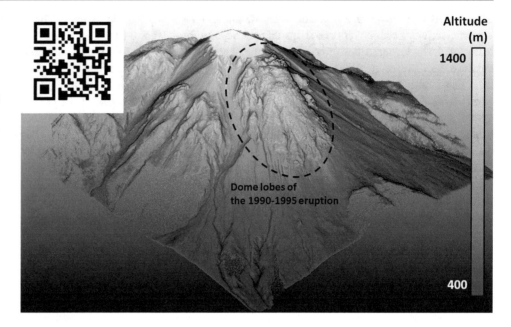

Fig. 6.4 Vertical topographic change at the dacitic dome between
2003 and 2015. The blue colour is coding the subsidence during this
period, while in red and orange, the deposition of 1–2 m and above 2 m
thickness is recorded. As these values compare two dates at the same
locations, movement and collapse of large rocks show important local
variations, but it is sometimes just the movement of these rocks (in the
yellow circle) that is recorded, rather than sudden localized erosion and
deposition. The gullies around the dome show that some of them are
being filled (temporarily) by material moving from the dome, while
others are being eroded further. The circled areas in yellow show blocks
that have moved or are difficult to define due to sub-vertical walls

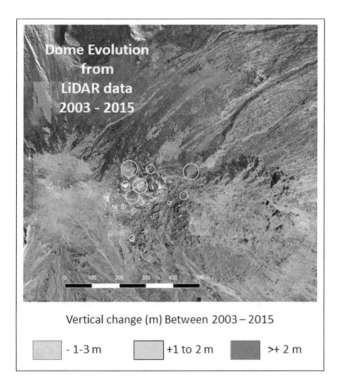

deposit, velocities of 5–10 m/s during deposition were
inferred from the TLS data [22] using a Riegl LMS-Z420i.

When investigating ignimbrites related to larger-scale
events, ALS is preferred over TLS. At Volcano Mazama, that
erupted around 5 million years ago, a series of ignimbrites
that are the results of a caldera-forming eruption were
investigated combining field survey and LiDAR data analysis
from a dataset of more than 5 billion data points [36, 37] .
Robinson et al. [37] used the shape, orientation and composition of ignimbrite benches and ridges and hummocky features to infer the origin and the directions of the different
deposits. Despite the relatively long time-span since the
eruption, the authors could even detect secondary collapse
scarps in the like of the ones observed at Mt. Pinatubo and
small collapse features parallel to the secondary pyroclastic
flow directions they named "scallops" (Fig. 8, p. 73). For
large-scale ignimbrites that can mantle a whole landscape,
making sense of those features from fieldwork alone would
be an extremely long process, and forested land does not
allow for traditional photo-interpretation. It is therefore only
the combination of fieldwork with LiDAR data that provides
the possibility to volcanic geomorphologists to explain the
features at such scale.

Large-scale ignimbrites can in turn also show sub-vertical
features that are difficult to map from the ALS alone, and it
is often accompanied by field GNSS and other ground-based
data acquisition or replaced by TLS when feasible. Faulting
in eroded volcanic tuffs (the Bishop Tuff in the Volcanic
Tableland of California), for instance, combined LiDAR and
GNSS RTK [11], while rockfall hazards identification from
steep ignimbrite faces preferred the use of TLS [46].

Unconfined lava flows, deposits and volcanic apron—As volcanic activity occurs over time-spans that are beyond the reach of a single generation, one of the approaches is to measure deposits of previously erupted material, such as lava flows and deposits on volcanic aprons. This has a particular influence because it is often on those volcanic aprons and eventually fans that communities develop. From a hazards and disaster risk management perspective, such measures also provide information on the potential behaviour of a volcanic structure and then the related hazards. From this perspective, Turner et al. [43] have investigated the potential of UAV photogrammetry (using a fix-wing SenseFly WingletCAM with a Canon IXUS 127 HS camera system telemetered with ground GNSS and GCPs) to foresee the potential paths of pahoehoe lava flows. As the great majority of physics-based lava flow model results are strongly controlled by the quality of the DEM, accurate and high-resolution DEM data is essential to predict the pathway lava will take [43]. The authors further explained that integrated RTK technology to UAVs is also key to further development, as it can be dangerous to go and deploy GCPs in hazardous areas, as we have already mentioned for in-vent monitoring. Such monitoring is further valuable as it provides numerical values of the volumes being erupted. For instance, it has been calculated at the Rancho Seco Monogenetic Volcano that eruption durations are typically 2–6 years, with inferred effusion rates of 4–15 m^3/s based on existing lava flow deposits. These measures were performed using LiDAR-based DEM [35].

6.2.4 Volcanic Micromorphology

Difficulties	– Small-scale vertical variations – In a climate with important moisture, the return of vegetation can be very quick
Recommended tools (not exhaustive)	– TLS – Hand-held cameras photography for photogrammetry – Small UAV photography operated over a very short range

At a smaller scale, point-clouds can help resolve features such as the network and opening of cracks in lava flows [31] as well as their length density for instance. It is also an important ally to determine the roughness of volcanic outcrop [17] and can even be used to measure density of granular unconsolidated volcaniclastic sediments, which are famously difficult to measure due to their unwelded and heterometric characteristics.

Surface erosion on ignimbrites and lahars, which can be unwelded and are deposits of heterogeneous material will often lead to the transport of the fines, while larger coarse sand and gravels can remain in place, eventually creating small-scale columns protected by a larger gravel, rock (Fig. 6.5).

On top of other measurements, a deep fracture of about 33 m deep and 28 m wide was measured in the andesitic dome [5], a feature that appeared as the dome contracted in the aftermath of the eruption.

Features that are studied with often limited topographic range also greatly benefited from the advent of high-resolution topography. Dering et al. [7] wrote about dike studies that "[…] our ability to develop new insights is limited by the continuity of high-precision data that can be extracted from the field. The emergence of Unmanned Aerial Vehicles (UAVs, a.k.a. drones), photogrammetry and other sensor techniques are set to greatly improve our field capabilities by collecting data at higher density and with greater precision". He presented a hypothesis-driven workflow to use UAV photogrammetry and then SfM that starts with the formulation of a hypothesis, to then define the aerial extent of the survey as well as the minimum number of GCPs and the right camera distance to the object to obtain the desired resolution. From this setup, the number of images, its geo-referencing and the flying patterns can be defined to perform the survey [7]. He further defines the division between fixed-wing, multirotor and terrestrial platforms depending on the working scale, to determine Breccia, the presence of intrusive contact, the trace of fracture, measure of vesicle and phenocryst, naturally performing a zoom on the feature. And this is arguably one of the main strengths of SfM-MVS, and it is its ability to work at variable scales within camera recording ranges and as long as a reference framework can be defined. For instance, microscope cameras are ill-suited for the technique. Finally, the author also demonstrates that the ability to gain photographs of the surface and the 3D data is essential to calculate dyke dipping, the walls around the dike, i.e. the 3D geometry of the intrusion.

6.2.5 4D Measurement of Lava in Motion

All the landforms and deposits we depicted in the sub-sections above can all be considered as "static" at the time-scale at which the photographs or laser scans have been taken. But, the speed at which laser scans are acquired and the capacity to use several cameras together to capture a scene in 3D has also paved the way to acquire time-lapse series of a moving object, even when the latter is a flow. We are therefore observing a densification of the dataset both in space and in time. Indeed, we moved from a quest to capture one or a few hundred precise and accurate points (with DGPS and later on GNSS technology either RTK GNSS or PPK GNSS) to the acquisition of several million points on a static scene. After this spatial densification of the point-cloud, we are now experiencing a time densification [20]. The same

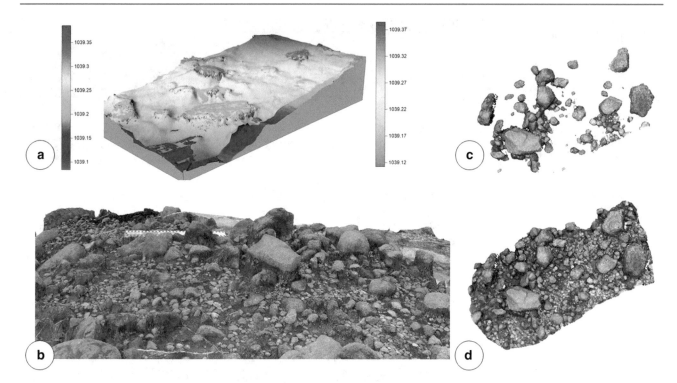

Fig. 6.5 Surface of the 2010 ignimbrites at Merapi Volcano in 2013, after 3 years of rainfalls (measurement carried out with a Canon EOS Kiss M2 camera, controlled with a set of rulers on the ground). **a** Elevation of the supposed original topography, and the present topography underneath (mostly invisible); **b** 3D view of the pointcloud with RGB mapping on each point from the photographs; **c** top of the topography and location where the base of the previous topography was assessed to have been; **d** present-day topography showing the same location as **c**

authors had already shown the possibility to capture the velocity magnitude of the flow from imagery at Mt. Etna in Sicily (Italy), using a single camera that is displaced. They had recorded maximum velocities of 0.9–1.5 m/min for cross-sections with a longitudinal slope of 22–34° [21].

Finally, before I draw you into the next section with worked examples, I should mention that although the field has seen amazing technological development, further growth is presently occurring, by notably combining underwater data and terrestrial data for volcanoes. Walter et al. [45] used underwater imaging and UAV photogrammetry to gain an insight of a geyser vent in a geothermal field of Iceland for instance.

6.3 Worked Examples with CloudCompare and Autodesk Recap

6.3.1 Finding, Loading and Doing Virtual Fieldwork on a US Volcano: Mt. Shasta of California

If you work in Northern America, Europe, Japan, Taiwan, etc., although some countries will have restrictions on who can access and where from, there is often a wealth of open-source LiDAR data, and often even so for volcanoes as

there is a great chance other scientists have also generated other point-cloud datasets, which they will be very happy to share, if they are properly acknowledged or eventually associated to your research (ethical data usage and sharing is essential). In this first example, we are using the USGS database of topographic LiDAR (Fig. 6.6a) from which you can download the topographic LiDAR data for Shasta Volcano for instance. From the online tool of the USGS, you can download the point-cloud in LAZ format, which can be opened seamlessly in CloudCompare (Fig. 6.6b), and unless you are interested in the intricacies and details of the topography, a set of sub-sampled point-clouds will work much faster. To sub-sample your point-cloud, the easiest tool is certainly by using CloudCompare, in which you select your point-cloud (not the folder, but the point-cloud under the folder) and you then just click the sub-sampling button (in the present version, a logo with a set of blue and red points), then you select the modalities of sub-sampling (for instance, in space with a minimum point separation of 1 m), and that's it, CloudCompare will operate its magic for you (Fig. 6.6c, d).

Finally, for basic measures on the point-cloud, you can use CloudCompare, but I am presenting here another tool that is free to use for students and academics and researchers at higher-education institutions: Autodesk Recap. In

Fig. 6.6 Basic steps from importing the LiDAR data of the summit of Mt. Shasta (North California, USA) to make linear measurements on the topography as a virtual terrain. **a** Download page of the USGS and that can be found in your favourite search engine typing USGS and LiDAR; **b** the set of downloaded LiDAR tiles (a total of 18) coloured by altitude between 2700 and 3500 m using the standard rainbow set of colour in CloudCompare; **c** zoom on one of the tiles in CloudCompare with the return intensity being the source of differentiation; **d** the same tile as in "**c**" but altitude coded and with the original point-cloud decimated to a 2 m interval; **e** another tile from the Shasta summit was imported into the Autodesk Recap software, where various sets of linear measurements and angle measurements are possible, as well as the use of annotations. The pointcloud with the annotations can be shared for collaborative work, teaching and learning

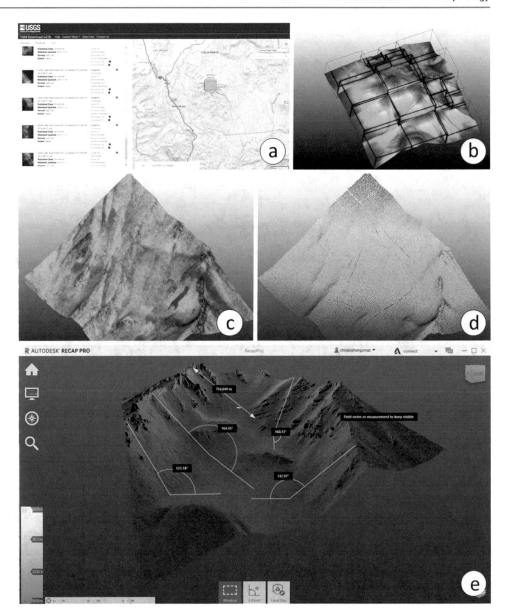

Autodesk Recap, you can load one or several of your point-clouds, make several calculations, such as the length or the angle between two segments on a surface, and you can also add annotations on the point-cloud, which can be all shared with others in 3D, so that you can use it as a collaboration tool as well (Fig. 6.6e).

6.3.2 Measuring the Depth, Eroded Bulk Volume (Material and Air/Water in the Volume) and Erosion Rate from UAV-Based Photogrammetry at Izu-Oshima Volcano

In the present example, we are interested in measuring the erosion rills at the surface of a slope on the East

flank of Izu-Oshima Volcano, in an area of 114,000 m^2 (400 m $\times$ 285 m) at the limit of the "desert" located inside the caldera.

A. Planning and Preparation

If you are doing a student project or later on in your career working on a geomorphologic project, the crucial first step is planning. You need to first develop a good understanding of what it is that you want to show, how to collect your data (extent, resolution, etc.) and what are the necessary elements you need to measure the reliability/accuracy of the data.

When you do your fieldwork, you do not always have access to the best equipment nor all the equipment you wish you had. I started this work just after joining my present position at Kobe University, and as there was no laboratory nor any equipment of any kind, I had to buy or rent all the basic material, so that I had to renounce a high-precision GPS or GNSS for differential GNSS or GPS. For this survey, I therefore had a basic total station with a range of 300–350 m and a phantom 4.

B. Data Acquisition

To create the set of control points, I therefore placed the UAV at point 5 and recorded the position using the UAV GNSS, which is not providing me with an absolute precision better than 1–5 m, but as I was interested in the relative precision, I settled for it (later on, I also used an existing LiDAR dataset, but let's assume that you don't have access to such a good dataset). Then, I placed the total station (at the location of point 5) and recorded the location of the other control points (making sure that the total station was properly oriented along the geodetic direction). As I only combined the UAV data and the total station in the processing phase, I did not enter the GPS data in the total station and calculated the position of each point afterwards. After creating a table (CSV format), I loaded them in Metashape-Pro and placed 7 of the 20 points on the photographs, in order to constrain the dataset. Please note that although it yielded good results, the limited number of points remaining to control the quality of the point-cloud was relatively limited and was not sufficient to generate a statistically significant set of data to differentiate by location of quality of the point-cloud. I am sure that the reader will not make the same mistake and collect more GCPs and more check-points for the error analysis. About 3–12 cm error.

C. Data Processing: From Point-Cloud Generation to DEM

Once the photographs have been acquired, they serve as input to the Metashape software, in which a set of tie points was added before the sparse pointcloud generation, to (1) ease the process and make the calculation faster and (2) to constrain eventual errors. Overall, try to spread your tie points as much as possible on the surface you want to map. Then, the method SfM-MVS itself was calculated from 150 photographs taken at ~ 75, ~ 50 and ~ 25 m above the ground surface (i.e. between 590 and 556 m a.s.l.). Then, I simply followed the different steps in the software to go from sparse pointcloud, to dense pointcloud, to mesh, DEM and orthophotograph. At this stage, you want to investigate the quality of your pointcloud and you can follow the methods provided in Chap. 3 to calculate the error and understand how meaningful your measurements truly are (Fig. 6.7).

D. Surface and Volume Calculation

Once you have cleaned your point-cloud from all the unwanted elements (it can be noise in the point-cloud) or the vegetation (in the present case, the spectral signature even in RGB is very different, so that it is possible to do it automatically, but this process may require a lot of hands-on-deck time), you are ready to compute the volume underneath your point-cloud. This measure can be done rapidly in CloudCompare by using the following tool:

> Tools > Volume > Compute 2.5D Volume

This tool creates a pseudo-grid over your pointcloud that is placed "flat" on a $Z = 0$ plane, so that you get a record of the column underneath the dataset. There are several ways to calculate the volume for each grid cell (see the point-cloud processing chapter for more details), and you can choose to use the mean elevation of the points present in the cell, the maximum elevation or the minimum elevation. For a LiDAR dataset that has the vegetation and the ground, this can be useful to calculate the volume occupied by the vegetation above the ground (let us say for under-canopy meteorology for instance). Each of the three methods available for the tool in CloudCompare (and you can very well use other calculation using the LidR package in R, which let you decide the values you want to assign to each grid cell) will return slightly different results depending on the size of the grid cell used (Fig. 6.8). The larger the grid cell, the more topographic variation is included in one cell and the larger the difference between the different methods. It is therefore important to match the size of the grid cell to the features you are trying to measure: measuring volumes of a gully that is 20 cm wide with a 1 m grid cell division may not yield the results you are expecting for instance.

To avoid using a grid to measure the volume, you can also do it from the meshed point-cloud, by measuring the surface of the triangles that constitute the mesh and the volume it contains (it is not the volume to a 0 level, but the volume contained). To do so in CloudCompare, you will select the portion of the point-cloud you are interested in (in the present case, we have a gully with two branches that extend over 100 m in length, for a total length of 125 m, adding the lengths of the two branches – you can calculate manually the distance from the point-picking tool of CloudCompare).

To extract the zone of interest, you need to segment your point-cloud, by using the "segment" tool (a pair of scissors in CloudCompare) and I have designed it as a pseudo-watershed around the gully I am interested in, so that

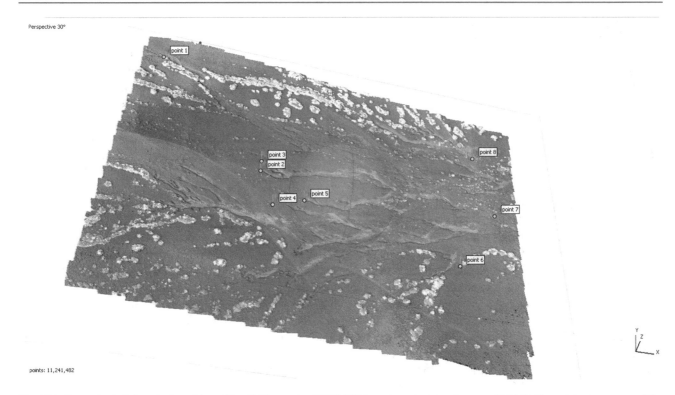

Fig. 6.7 Reconstructed slope in the caldera of Izu-Oshima, using SfM-MVS from imagery taken by a small UAV. The points show some of the GCPs used to check for the correctness of the dataset against existing LiDAR data and data collected using a total station

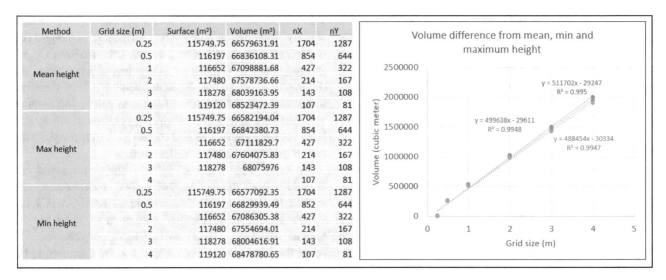

Method	Grid size (m)	Surface (m²)	Volume (m³)	nX	nY
Mean height	0.25	115749.75	66579631.91	1704	1287
	0.5	116197	66836108.31	854	644
	1	116652	67098881.68	427	322
	2	117480	67578736.66	214	167
	3	118278	68039163.95	143	108
	4	119120	68523472.39	107	81
Max height	0.25	115749.75	66582194.04	1704	1287
	0.5	116197	66842380.73	854	644
	1	116652	67111829.7	427	322
	2	117480	67604075.83	214	167
	3	118278	68075976	143	108
	4			107	81
Min height	0.25	115749.75	66577092.35	1704	1287
	0.5	116197	66829939.49	852	644
	1	116652	67086305.38	427	322
	2	117480	67554694.01	214	167
	3	118278	68004616.91	143	108
	4	119120	68478780.65	107	81

Fig. 6.8 Differentiation of the calculated volumes based on the grid cell size and the chosen algorithm

I have a portion of the topography I assumed has not be changed by erosion (in blue in Fig. 6.9), as well as the gully, where erosion occurred (in red in Fig. 6.9). Once you have segmented the area of interest, duplicate your zone of interest and use the segment tool to separate the gully from the surrounding area. Once this is done, you can mesh the point-cloud without the gully at its centre, so that the triangles of the mesh will be interpolated from each banks of

the gully, not taking into account the eroded area, and then repeat the same process on the point-cloud that contains both the gully and the surrounding area.

You now have one meshed data that is the pseudo-watershed without the gully, and the pseudo-watershed with the gully. Still in > tools > Mesh you can now use the volume calculator. In the present case, I calculated a meshed volume with the gullies of 9186.29 m³, and the meshed

Fig. 6.9 Eroded zone at Izu-Oshima, with the gully erosion in red and the surrounding non-eroded surface in blue. Interpolating the data without the gully and assuming that there was no gully at that location when the volcanic fallout material draped the area, you can then estimate the volume of material that has been eroded and thus the erosion rates when the eruption date is known

volume without the gullies is 8923.08 m^3. This results in a volume of approximately 263 m^3, which brings back to the total length of the gully is 2 m^2 of erosion (think of it as a cross section) per metre of gully. From the point-cloud of the gully only, you can then use the volume tool, which also gives you the surface (in the present case 1158 m^2, so that you can calculate the erosion to be 0.227 m).

The averaged erosion is therefore 22.7 cm/m^2, which can then yield an erosion rate since the material emplacement in 1986 of 0.7 cm per m^2 per year for the gullies. You would need to get the density of the gullies on the all surface to have an estimate of the gully erosion per unit surface. This result only tells you that the gullies have been eroding at a rate of 0.007 m/m^2/year.

E. Discussion of the Results: Erosion Rate (Meaningful or Not)

Leaving aside the issues of error analysis in the dataset, which is a recurring topic in the engineering literature, I would like the reader to reflect on the significance of the data and results obtained from a geomorphological perspective. As it is often the case when a new technology or a technology new to a field is introduced, there are two sets of human behaviours that can be observed that are either of the domain of fascination and admiration, as though the silver bullet had been invented, and there is another camp that just does not see the use of it and regret old-time geomorphology when it was done with a compass and other pocket tools while walking the hills and other landscapes. I don't think that there is anything wrong with any of these positions, as debates, and the ability to have divergent views is most certainly the main ingredient for a good scientific recipe (even if journals' strategies are often pulled in one direction or another based on the camp the chief editor is in), but there is certainly a need to strike a balance for the sake of geomorphology. Indeed, metric calculations of errors based on the accuracy of the data and the method proposed are all essential to be able to compare dataset taken by different operators, using different methods and techniques, but this error is often stated as the reasonable margin from which the geomorphologist can make sense of these data. I would argue that in an example like the one I am proposing as an example here, the error margin of the method/technology is eventually not essential in the understanding of the geomorphologic processes and their measurement.

I would like to start this argument by turning our attention to the 2013 heavy-rainfall event that struck Izu-Oshima and turned portions of the external slopes made from ejecta (much older than those of 1986) that mantled the slope and which were supposed to be relatively stable. This once in a century "landslide turned into lahar" event.

6.3.3 SfM-MVS for Laboratory-Scale Volcanic Structures and Forms

SfM-MVS can be also used in controlled environments to help modelling processes in flumes [32] and other inclined plans. SfM-MVS has a great advantage against other laser and laser sheets techniques, as it can be triggered from multiple cameras all at once together, so that processes can be measured from one side to the other of a flume or an experiment at the very exact same time. Each time that one uses a laser or any other form of height measuring method (even a rather rapid TLS), all the 3D points are created incrementally, so that a topographic surface measured with such method offers a view between time t and $t + \Delta t$, but not all at the same time. SfM-MVS can solve this issue. In other words, if you simulate a sediment transport process or any other process that will transport sediments and modify the "micro-landforms" at the time-scale of the measurement, you may want to consider a multiple camera set to capture all of your scene at the very same time.

The other advantage of working indoors is to limit unnecessary shades and shadows so that you have an optimum capacity. The element to keep in mind when shooting indoors is that the ceiling lights and other lighting features may create reflections at the surface of the water—if you are using water—in such a case will create errors and hamper your measurement. You therefore want to have diffuse light, a bit like the one that photographers use by illuminating a white reflector. For the experiment XP3 provided as an example underneath, I use normal light and a low-cost white bed linen (Fig. 6.10: there are no reflections from the ceiling lights thanks to a shade used here).

SfM-MVS is an accurate method when developed properly in the laboratory with the appropriate software. Morgan et al. [32] have tested the use of SfM-MVS in laboratory setting against two different terrestrial laser scanner data (Leica ScanStation and Faro Focus3D) using flumes between 0.22 and 6.71 m width and between 9.14 and 30.48 m lengths. The results between SfM-MVS and TLS are comparable, but their results also show that the software Photoscan was more appropriate than VisualSfM in the applications they presented.

SfM-MVS notably provides sufficient precision and accuracy to model elevation changes due to soil erosion with vertical changes between ~ 1 and ~ 80 mm, as it has been shown for experimental plots of 68 cm × 75 cm [1]. The authors made an important point: "the computed sediment yield was 13% greater than the measured sediment yield", showing that the surface data cannot be considered as

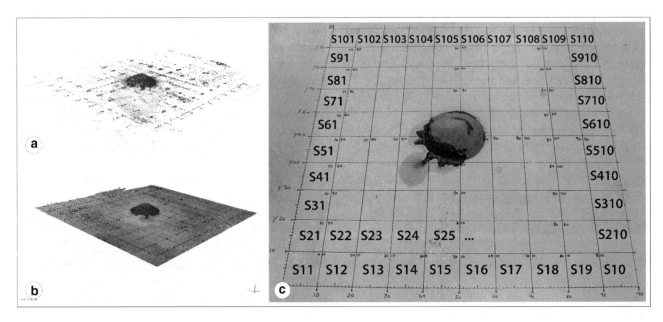

Fig. 6.10 Sand cone morphology and water effect experiment done at the Geography Department of the University of Canterbury in 2014. **a** Sparse point-cloud (28,628 pts) calculated using 121 ground control points (GCPs) on a grid, in a "difficult" environment for SfM-MVS as only the grid and marking can be used for object detection; **b** dense point (14,590,822 pts) calculated from the data in **a**; **c** photograph of

Experiment 3 (XP3) setup showing the sand pile constructed from (1) 300 ml uncompacted fine sand, to which 300 ml of water was added from a height of 10 cm above the top of the pile from a tube of half a centimetre diameter, on top of which 300 ml of sand was poured again, being total of 900 ml of water and material

Table 6.1 Statistical table of the values calculated from the sparse pointcloud generated on the laboratory floor

Name	Pointcloud	Number of points	Subset		RMSE(Z)	MAE(Z)	MDAE (Z)	Min.	Max	Median	Mean	1st Qu.	3rd Qu.
			Location	Pt nb.	[m]	[m]	[m]	[m]	[m]	[m]	[m]	[m]	[m]
XP3	Sparse	28,628	X < 0	523	0.0018675	0.00171	0.0017	0.0054	0.0012	0.0017	0.001625	0.0012	0.0021
			X > 1	320	0.00126	0.000751	0.0005	−0.01	0.0069	0.0004	0.0004	0.0001	0.0008
			Y < 0	404	0.001405	0.001027	0.0008	−0.0026	0.0068	0.0007	0.0008	−0.0001	0.0015
			Y > 1	11	0.00159	0.001263	0.0013	−0.0002	0.0034	0.0013	0.001227	0.00025	0.0018
			S11	396	0.00183	0.0016	0.0015	−0.0017	0.0092	0.0015	0.001638	0.0013	0.0019
			S12	134	0.00127	0.00091	0.0008	−0.003	0.0052	0.0007	0.00077	0.0002	0.001
			S13	142	0.001	0.00054	0.0004	−0.0071	0.0052	−0.0001	−0.00069	−0.0004	0.0002
			S14	116	0.00054	0.00047	0.0005	−0.0015	0.0009	−0.0004	−0.00032	−0.0006	−0.00017
			S15	206	0.00081	0.00054	0.0005	−0.0025	0.0063	−0.0003	−0.00031	−0.00057	−0.0001
			S16	62	0.00086	0.00061	0.0004	−0.0034	0.0008	−0.0004	−0.00047	−0.0007	−0.0002
			S17	58	0.00114	0.00074	0.0004	−0.0031	0.0041	−0.0001	−0.00002	−0.0005	0.0002
			S18	53	0.00081	0.00056	0.0004	−0.0031	0.0018	−0.0001	−0.00011	−0.0005	0.0004
			S19	53	0.001	0.00794	0.0008	−0.0012	0.0034	0.0007	0.000677	0.0002	0.001
			S10	44	0.00187	0.00136	0.0009	−0.0035	0.0064	0.0008	0.001043	0.0005	0.0014

The location Sn refers to the sets of sample squares delineated by four GCPs, as seen in Fig. 6.10c. In the table, Pt. nb. signifies number of points for the subset (note the little number of tie points detected due to the highly uniform concrete surface), RMSE(Z) is the root mean square error in Z, MAE(Z) is the mean absolute error in Z, MDAE(Z) is the median absolute error in Z; Min. is the minimum value, Max. is the maximum value, Median and Mean are self-explanatory, and 1st Qu. is the first Quantile and 3rd Qu. is the third quantile. The last 6 indicators provide estimates of the distribution of the points' elevations. The formula to calculate RMSE, MAE, MDAE, the mean and median is provided as a reminder in the post-processing section of this chapter

sufficient when working at high resolution, because there are other factors that come at play (see the chapter on the necessity to integrate sub-surface geotechnical considerations to improve the interpretation of HRT data) (Table 6.1).

When you are at this stage, you have a pointcloud and you have calculated error for your dataset following different methods that you can compare, the next potential step you can think about is the spatial distribution of your error, that is the result of both the spatial variability of the surface and also the spatial distribution of the GCPs. In this example, we are using a regular horizontal concrete surface with potential millimetre-scale imperfection, but overall regular. It is then a correct assumption to assume that the surface characteristic variations are not at the origin of the error, but it is the number of overlapping photographs and the number of GCPs. But, even with regularly spaced GCPs, error that persists will also show spatial variability (Fig. 6.11).

This is particularly important if you are working in the laboratory with SfM-MVS or laser technology, because the scale reduction usually applied to scaled models will yield increased importance to small-scale features and changes (what would be a 10 m scale feature in the environment, might only be a few millimetres in the laboratory). It is then essential to understand how your error may have varied across your experimental setup.

Arguing that these variations may also occur in the field, and that they may also be emphasized by the reality of the terrain, if you have the opportunity to work on those spatial variations with a sufficient number of GCPs, I would then advise you to try working on those variations as well, so that you can determine whether you can have the same levels of confidence across your all research field.

Conclusion

Volcanoes are complex structures that expand beyond the reach of surface investigation, but this surface is the window to the processes occurring underneath and to understanding the processes at the interface with atmospheric and other Earth surface processes.

After a rapid introduction to some ideas of volcanic geomorphology, the present chapter has brought you through a progressive zoom to volcanic-scale geomorphology, then slope geomorphology, feature-scale geomorphology (e.g. lava dome), to finish with micro-geomorphology on a pyroclastic flow deposit. After this progressive zoom, I provided you with worked examples at Izu-Oshima measuring erosion in the aftermath of the 1986 eruption and then to the laboratory, where a small conic feature representing a potential volcano was investigated. In the present case, we used the dataset to investigate the error distribution and how to deal with error in the laboratory.

From this chapter, I expect the students to have develop some understanding of the difficulties related to volcanic geomorphology, an understanding on how we can potentially

Fig. 6.11 Statistical descriptors of two sets of sampling quadrants, located at the top edge of the survey area (S101–S110) and between 10 and 20 cm from the lower limit (S21–S210). This displays the increasing variability (in this case error) generated by the SfM-MVS algorithm, proof of the importance of GCPs and the position of the GCPs

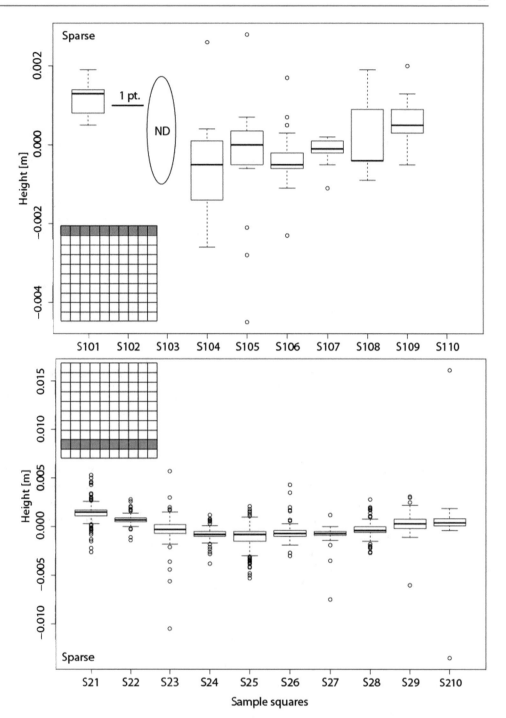

overcome those, and I hope that I could also tickle some of your "little grey-cells" (to use the words of A. Christie's character Hercule Poirot) and that you may have some new ideas to work on volcanoes with pointcloud technologies.

Problems and Potential Projects

It is most likely that you are reading those lines thinking, well I do not live near a volcano, what can I do as a project … Well that's when the magic of the Internet and laboratory-scale experiments come to the rescue.

1. Go to the OpenTopography.org website or the USGS website and choose one of your favourite volcanoes in The Cascade for instance, and download the LiDAR dataset. Because those volcanoes have been well studied (notably by the USGS), you can look at the walls of calderas and craters, or the walls of gullies used by lahars, as well as the pseudo-terraces and try to determine since the last lahar or pyroclastic flow or ash fallout event, how much material transfer and then erosion has

occurred? What does this not tell you about the rhythms of erosion?

2. Keeping your favourite volcano topographic data (from Project 1), downsample your pointcloud on a grid at 50 cm, 1 m, 2 m, 5 m, 10 m, 25 m, 30 m, 35 m and 50 m horizontally. Make a table of the features that can be identified at the different scales, and compare topographic profiles of selected features, but at different scales. Besides the loss of precision, what are the other negative effects that you can detect? What is the contamination of neighbouring topographies on the object you are investigating? How can you relate those to what can be seen or not from traditional SRTM for instance?

3. For this third project, let's leave our real-life volcano and go to the laboratory. Fill up a balloon with your favourite liquid (I mean water), using a long nozzle and clamp the end of the rubber tube, so that the liquid does not escape. Then, bury the balloon underneath wet sand. The wet sand should look like the pile I have used in Fig. 6.10. Use your camera and a setting similar to the one I used to calculate the volume of the volcano using SfM-MVS. Empty the balloon from the liquid, and watch a caldera form where your volcano was. USe SfM-MVS to measure the second volume of the volcano, and compare the loss of volume to the amount of liquid that was taken out of the balloon. How accurate is your prediction? What are the sources of error? Now try the same operation but the other way around, this time pump some liquid in the balloon. Obviously, the original topography is not recreated, but compare the injected liquid to the new volume of the volcano. How much error do you get? Try to understand what is the role of decompaction–recompaction (if you don't see anything, make sure you nicely compact your sand at the start of the experiment, for improved results).

4. Measure a volume of sand in a bucket, and poir it from a given height to form a conic volcano on the floor of your laboratory. Using SfM-MVS or a TLS, measure the volume of the cone. How much change in the volume can you detect? Then, pour a known volume of water over the volcano until it reaches a shield-volcano shape. Measure the volume of the volcano this time again, and if no water escaped from your experiment (you could collect flowing water in a recipient), measure again the volume of your new volcano. What can you say about the empty pores filled by water? How can you make this measurement while still taking into account the change in compaction? (make sure that you think over this process before starting the experiment, as you will need to do some "extra-measurements" involving the dry density of the material and not the bulk density provided by the 3D SfM-MVS or your laser scanner).

Acknowledgements Part of the data acquired and processed for this chapter was supported by research and collaboration funds from the Graduate School of Maritime Sciences of Kobe University and by travel funding from Universitas Muhammadiyah Surakarta in Indonesia. I would like to also thank the members of Universitas Muhammadiyah Surakarta who supported the research in Indonesia, especially Dr Aditya Saputra for his continued support and help in the field.

References

1. Balaguer-Puig M, Marques-Mateu A, Lema JL, Ibanez-Asensio S (2017) Estimation of small-scale soil erosion in laboratory experiments with structure from motion photogrammetry. Geomorphol 295:285–296
2. Bisson M, Spinetti C, Neri M, Bonforte A (2016) Mt. Etna volcano high-resolution topography: airborne LiDAR modelling validated by GPS data. Int J Digit Earth 9:710–732
3. Branney MJ, Kokelaar P (2002) Pyroclastic density currents and the sedimentation of ignimbrites. Geol Soc Mem 27:143 p
4. Csatho B, Schenk T, Kyle P, Wilson T, Krabill WB (2008) Airborne laser swath mapping of the summit of Erebus volcano, Antarctica: applications to geological mapping of a volcano. J Volcanol Geotherm Res 177:531–548
5. Darmawan H, Walter TR, Sri Brotopuspito K, Subandriyo, Agung Nandaka IGM (2018) Morphological and structural changes at the Merapi lava dome monitored in 2012–2015 using unmanned aerial vehicles (UAVs). J Volcanol Geotherm Res 349:256–267
6. De Beni E, Cantarero M, Messina A (2019) UAVs for volcano monitoring: a new approach applied on an active lava flow on Mt. Etna (Italy), during the 27 February–02 March 2017 eruption. J Volcanol Geotherm Res 369:250–262
7. Dering GM, Micklethwaite S, Thiele ST, Vollgger SA, Cruden AR (2019) Review of drones, photogrammetry and emerging sensor technology for the study of dykes: best practises and future potential. J Volcanol Geotherm Res 373:148–166
8. Dibacto S, Lahitte P, Karatson D, Hencz M, Szakavs A, Biro T, Kovacs I, Veres D (2020) Growth and erosion rates of the East Carpathians volcanoes constrained by numerical models: tectonic and climatic implications. Geomorphology 368:107352, 1–21
9. Favalli M, Fornaciai A, Pareschi MT (2009a) LIDAR strip adjustment: application to volcanic areas. Geomorphology 111:123–135
10. Favalli M, Karatson D, Mazzarini F, Pareschi MT, Boschi E (2009b) Morphometry of scoria cones located on a volcano flank: a case study from Mt. Etna (Italy), based on high-resolution LiDAR data. J Volcanol Geotherm Res 186:320–330
11. Ferrill DA, Morris AP, McGinnis RN, Smart KJ, Watson-Morris MJ, Wigginton SS (2008) Observations on normal-fault scarp morphology and fault system evolution of the Bishop Tuff in the Volcanic Tableland, Owens Valley, California, U.S.A. Lithosphere 8–3:238–253
12. Fornaciai A, Bisson M, Landi P, Mazzarini F, Pareschi MT (2010) A LiDAR survey of Stromboli volcano (Italy): digital elevation model-based geomorphology and intensity analysis. Int J Remote Sens 31:3177–3194
13. Gomez C (2012) Multi-scale topographic analysis of Merbabu and Merapi volcanoes using wavelet decomposition. Environ Earth Sci 67:1423–1430
14. Gomez C, Janin M, Lavigne F, Gertisser R, Charbonnier S, Lahitte P, Hadmoko SR, Fort M, Wassmer P, Degroot V (2010) Borobudur, a basin under volcanic influence: 361,000 BP to present. J Volcanol Geotherm Res 196(3–4):245–264
15. Gomez C, Kennedy B (2017) Capturing volcanic plumes in 3D with UAV-based photogrammetry at Yasur Volcano—Vanuatu. J Volcanol Geotherm Res 350:84–88

16. Gomez C, Hayakawa Y, Obanawa H (2015) A study of Japanese landscapes using structure from motion derived DSMs and DEMs based on historical aerial photographs: new opportunities for vegetation monitoring and diachronic geomorphology. Geomorphology 242:11–20

17. Gomez C, Kataoka K, Saputra A, Wassmer P, Urabe A, Morgenroth J, Kato A (2017) Photogrammetry-based texture analysis of a volcaniclastic outcrop-peel: low-cost alternative to TLS and automation potentialities using Haar wavelet and spatial-analysis algorithms. Forum Geogr 31. https://doi.org/10.23917/forgeo.v3lil.3977

18. Grosse P, van Wyk de Vries B, Euillades PA, Kervyn M, Petrinovic IA (2012) Systematic morphometric characterization of volcanic edifices using digital elevation models. Geomorphology 136:114–131

19. Hanagan C, La Femina PC, Rodgers M (2020) Changes in crater morphology associated with volcanic activity at Telica volcano, Nicaragua. Geochem Geophys Geosyst 15 p. https://doi.org/10.1029/2019GC088889

20. James MR, Robson S (2014) Sequential digital elevation models of active lava flows from ground-based stereo time-lapse imagery. ISPRS J Photogramm Remote Sens 97:160–170

21. James MR, Pinkerton H, Robson S (2007) Image-based measurement of flux variation in distal regions of active lava flows. Geochem Geophys Geosyst 8(3):Q03006, 1–16

22. Jessop DE, Kelfoun K, Labazuy P, Mangeney A, Roche O, Tillier JL, Trouillet M, Thibault G (2012) LiDAR derived morphology of the 1993 Lascar pyroclastic flow deposits, and implication for flow dynamics and rheology. J Volcanol Geoth Res 245:81–97

23. Jones LK, Kyle PR, Oppenheimer C, Frechette JD, Okal MH (2015) Terrestrial laser scanning observations of geomorphic changes and varying lava lake levels at Erebus volcano, Antarctica. J Volcanol Geotherm Res 295:43–54

24. Karatson D, Telbisz T, Worner G (2012) Erosion rates and erosion patterns of Neogene to Quaternary stratovolcanoes in the Western Cordillera of the Central Andes: an SRTM DEM based analysis. Geomorphology 139–140:122–135

25. Karlstrom L, Richardson PW, O'Hara D, Ebmeier SK (2018) Magmatic landscape construction. J Geophys Res Earth Surf. https://doi.org/10.1029/2017JF004369

26. Kereszturi G, Schaefer LN, Schleiffarth WK, Procter J, Pullanagari RR, Mead S, Kennedy B (2018) Integrating airborne hyperspectral imagery and LiDAR for volcano mapping and monitoring through image classification. Int J Appl Earth Obs Geoinf 73:323–339

27. Lahitte P, Samper A, Quidelleur X (2012) DEM-based reconstruction of southern Basse-Terre volcanoes (Guadeloupe archipelago, FWI): contribution to the Lesser Antilles Arc construction rates and magma production. Geomorphology 136:148–164

28. Lavigne F, Thouret J-C (2003) Sediment transportation and deposition by rain-triggered lahars at Merapi Volcano, Central Java, Indonesia. Geomorphology 49:45–69

29. Leverington DW, Teller JT, Mann JD (2002) A GIS method for reconstruction of late Quaternary landscapes from isobase data and modern topography. Comput Geosci 28:631–639

30. Major JJ, Pierson TC, Dinehart RL, Costa JE (2000) Sediment yield following severe volcanic disturbance—a two decades perspective from Mount St Helens. Geology 28:819–822

31. Massiot C, Nicol A, Townend J, McNamara DD, Garcia-Selles D, Conway CE, Archibald G (2017) Quantitative geometric description of fracture systems in an andesite lava flow using terrestrial laser scanner data. J Volcanol Geotherm Res 341:315–331

32. Morgan JA, Brogan DJ, Nelson PA (2017) Application of structure-from-motion photogrammetry in laboratory flumes. Geomorphology 276:125–143

33. Pesci A, Fabris M, Conforti D, Loddo F, Baldi P, Anzidei M (2007) Integration of ground-based laser scanner and aerial digital photogrammetry for topographic modelling of Vesuvio volcano. J Volcanol Geotherm Res 162:123–138

34. Pesci A, Teza G, Casula G, Loddo F, De Martino P, Dolce M, Obrizzo F, Pingue F (2011) Multitemporal laser scanner-based observation of the Mt. Vesuvius crater: characterization of overall geometry and recognition of landslide events. ISPRS J Photogramm Remote Sens 66:327–336

35. Ramirez-Uribe I, Siebe C, Chevrel MO, Fisher CT (2021) Rancho Seco monogenetic volcano (Michoacan, Mexico): petrogenesis and lava flow emplacement based on LiDAR images. J Volcanol Geotherm Res 107169, 1–19

36. Robinson (2012) High-resolution digital elevation dataset for Crater Lake National Park and vicinity, Oregon, based on LiDAR survey of August–September 2010 and bathymetric survey of July 2000. U.S. Geol Surv Data Ser 716. https://doi.org/10.3133/ds716

37. Robinson JE, Bacon CR, Major JJ, Wright HM, Wallance JW (2017) Surface morphology of calderaforming eruption deposits revealed by lidar mapping of Crater Lake National park, Oregon–Implications for deposition and surface modification. J Volcanol Geoth Res 345:61–78

38. Sakagami M, Sasaki H, Mukoyama S (2010) Estimation of ground movement caused by the 2000 eruption of Usu Volcano, from the geomorphic image analysis of LiDAR DEMs. Netw World Remote Sens 38:293–297

39. Sakuma S, Kajiwara T, Nakada S, Uto K, Shimizu H (2008) Drilling and logging results of USDP-4—penetration into the volcanic conduit of Unzen Volcano, Japan. J Volcanol Geotherm Res 175(1–2):1–12

40. Suwa H, Yamakoshi T (1999) Sediment discharge by storm runoff at volcanic torrents affected by eruption. Z Geomorphol Suppl 114:63–88

41. Tarolli P (2014) High-resolution topography for understanding earth surface processes: opportunities and challenges. Geomorphology 216:295–312

42. Thouret J-C, Oehler J-F, Gupta A, Solikhin A, Procter JN (2014) Erosion and aggradation on persistently active volcanoes—a case study from Semeru Volcano, Indonesia. Bull Volcanol 76(857):1–26

43. Turner NR, Perroy RL, Hon K (2017) Lava flow hazard prediction and monitoring with UAS: a case study from the 2014–2015 Pahoa lava flow crisis, Hawai'i. J Appl Volcanol 6(17):1–11

44. Walter TR, Salzer J, Varley N, Navarro C, Arambula-Mendoza R, Vargas-Bracamontes D (2018) Localized and distributed erosion triggered by the 2015 Hurricane Patricia investigated by repeated drone surveys and time lapse cameras at Volcan de Colima, Mexico. Geomorphology 319:186–198

45. Walter TR, Jousset P, Allahbakhshi M, Witt T, Gudmundsson MT, Hersir GP (2020) Underwater and drone based photogrammetry reveals structural control at Geysir geothermal field in Iceland. J Volcanol Geotherm Res 391:106282, 1–9

46. Wang X, Frattini P, Stead D, Sun J, Liu H, Valagussa A, Li L (2020) Dynamic rockfall risk analysis. Eng Geol 272:105622, 1–11

47. Zarazua-Carbajal MC, De la Cruz-Reyna S (2020) Morphochronology of monogenetic scoria cones from their level contour curves. Applications to the Chichinautzin monogenetic field, Central Mexico. J Volcanol Geotherm Res 407:107093, 1–17

Discussion and Novel Opportunities

7

Abstract

The last chapter builds on the previous 6 chapters, which are more technical, to propose a reflection of what it means for the geomorphologist to work with high-resolution pointcloud technology. First, the chapter explains that values such as sediment density and compaction, which could be safely ignored in the past are now having an increasing importance in the measured changes, because we can access minute variation, which have more origins than simply erosion and/or deposition. Secondly, the chapter explains that pointcloud technologies have generated a transfer of the geomorphological act, where instead of engaging a reflection leading to a measure in the field, pointcloud technologies are now bringing the geomorphological act in the virtual space. It is now in front of the computer that the geomorphologist will eventually chose a point or extract a transect or generate a 3D volume. The field can be fully automated, robotic almost. Finally, the chapter also presents new fields of opportunities that the geomorphologist can open with pointcloud technologies, like measuring density and density variations, compaction and decompaction occurring with Earth surface processes. These otherwise punctual measures that were limited to civil engineering use can now make their way to geomorphology as well. As a closing chapter, one can see that pointcloud technology is not only a technological change, it has modified the geomorphological act as well as its prospect and assumptions and paradigms.

Learning Outcomes: New Tools, New Paradigms and Space for Growth

After reading this chapter, the reader should:

– Be able to understand the limitations of present geomorphological paradigms when using pointcloud technologies and know what steps may need to be developed to overcome those limitations.
– Know how to measure bulk soil density using the advantages brought by pointcloud technologies.
– Understand why pointcloud technology generates a paradigm shift in geomorphology.
– Know how the methods allow to close both the spatial and the temporal gap in measurements, so that we can approach geomorphological reality even further.

Objectives and Content

The main objective of this chapter is to offer a reflection space on pointcloud technologies in geomorphology for geomorphologist. The reason that drove this choice is a career-long bitter-sweet taste that arose from my Ph.D. At the time I was taking my Ph.D., I received a scholarship from the French Government, which is for 3 years, and thanks to this scheme, I was receiving a salary and was given the chance to teach at university. And although this sounds like a dream spot for a Ph.D. student, I also felt limited by the 3 years period, as it did not give me sufficient time (maybe I am just slow) to reflect on the contribution my research was providing to the broader field of geomorphology, science and human knowledge as a whole. Obviously, a chapter is not enough to even start a conversation on those questions, but I thought that it was a good space to share some reflections and also concrete applications of pointcloud technology, which, I think, have a potential to be developed further.

Through this myriad ("point-cloud") of potential topics, my objectives are to engage the reader in one of the question, which I think merits a little space here, it is the question of what pointcloud have changed in the geomorphologic reflection. This is almost an epistemology question that looks at our practices yesterday and how our practices were informed by geomorphologic theories, and then now that we have pointcloud technology what are the changes that this technology brings. We can also ask whether past paradigm

C. Gomez, *Point Cloud Technologies for Geomorphologists*,
Springer Textbooks in Earth Sciences, Geography and Environment,
https://doi.org/10.1007/978-3-031-10975-1_7

and ideas of geomorphology are still fully valid with the high-resolution-topographic (HRT) data. Of course, it is bringing changes, because we need to now consider variables that were negligible when we were working at a broader scale, and now we do have to take them into account. It would be of little interest without a concrete example, I therefore start from a theoretical example of erosion deposition (close to the problem I am stating in the introduction), and I propose a concrete way to solve the issue, still using pointcloud technology. Namely, I introduce the—otherwise well-known in civil engineering and geotechnical engineering—issue of soil compaction and how we can measure it near the surface and on outcrop with pointcloud technologies, and how this measure can lead to measures of density, which are then essential to correctly calculate erosion deposition.

Finally, this chapter acts as a closure to the first chapter. In the first chapter, we presented some of the changes in both concepts and opportunities offered by pointcloud technologies to geomorphologist, and in this chapter, we are going back to some of these ideas, and this is the second reason why I am presenting you with the result of some of the research in measuring soil bulk density with SfM-MVS. There is also behind this idea the further objective to be able to look at the energy distribution over a deposit (a.k.a. how much energy was available to pack one material of a given size at a given location, based on a given type of process).

Who is this chapter for and not for? If I had to pick one chapter, I think, I would need to be read—if not all the rest in this book—this would be the one. The present chapter is not a technical chapter that will help you get started with the point-cloud technologies; it is a chapter aimed at considering the role of point-cloud technologies in geomorphology and what is still missing in the geomorphological framework to fully integrated HRT point-cloud technologies. This chapter should interest those who want to develop further the field of geomorphology beyond the technical issues and concerns we are now focusing on (and this should not be seen as a critic, it is a natural step, but this is not where geomorphologists should stop).

What is the chapter about (in a few words)? This chapter is about the paradigm change related to HRT point-clouds in geomorphology and what are the adjustment necessary to further integrate this tool to the sciences. The chapter is concentrated on the question of how representative is a surface measurement at high resolution (as defined in HRT) and what sense does it make for geomorphology. From those considerations, it appears that elements of geotechnical engineering and science need to percolate through the discipline barrier, in order to explain the data

obtained from point-cloud HRTs. This needs for adjustment also emphasizes the fact that geomorphologists may not be measuring what they think they are measuring, and rather than being a negative, this may be opening the way for new forms of questioning in geomorphology.

7.1   Introductory Comments

As it has been described in the previous chapters, the last two or three decades have seen the rise in the field of geomorphology of very effective ways to acquire large amount of geomorphological data. Furthermore, the dataset can provide precise and accurate data to the centimetre and eventually millimetre scale and even so for spaces extending along tens to thousands of metres. In the long run, there is no question that the integration of those tools is marking a turn in geomorphology.

This being stated, the present chapter investigates what is still missing to make those techniques part of geomorphology fully. So far, we are still at the stage where scientists are applying the methods with different levels of success and accuracy, with the most "engineer"-oriented groups of geomorphologists and those coming from the fields of surveying, attempting to provide best-practice guidelines and protocols. Those guidelines and protocols are arguably essential from an accuracy and precision perspective, but it is only the accuracy and the precision of the measurement made.

HRT topography, in the same way than a 1/25,000 topographic map does, is a set of discrete measurements of the limit of an object, here the ground surface, and the atmosphere (it is bathymetry when the atmosphere is replaced by water). The concepts of geomorphology that use the land surface as a proxy of land surface and endogenous processes, and which also allow, through the description of a landform at a combination of scales, the definition of past processes is a paradigm developed before the advent of HRT topography. It is therefore important to ask whether the assumptions made from data collected at a given scale are still valid with HRT—please note that I am not using the terms of smaller or larger scale, because HRT is also breaking free from some of the barriers developed by those concepts. Indeed, in Chap. 1 of Fundamentals of Geomorphology, Huggett writes that "Landforms are conspicuous features of the Earth and occur everywhere. They range in size from molehills to mountains to major tectonic plates […]" (p. 3), and what HRT point-cloud technology has given is the ability to measure the molehills landforms at the spatial scales of mountains and eventually mountain ranges. Consequently, the division between small scale and large scale does not really exist anymore. When I was an undergraduate at Sorbonne University, I worked for my final undergraduate year on the lahars of Semeru Volcano, and it was possible to either do local topographic measurements at the decimetre to centimetre scale of the lahar pseudo-terraces, or it was possible to measure the volcanic shape at a 1/25,000 scale, but I would not have thought of investigating the shape of clasts at the surface of the lahar deposit for the entire volcano. The spatial scale and the scale of the object that could be investigated was directly related. The other important scale relationship that HRT point-clouds allow to break free from is the human-scale condition to measurement. I would not argue that measurement has totally broken free of the human-scale, but some of the human limitations have been overcome. Moreover, some of the scale limitations are not determined by the tools but by the scientific studies design choices. For instance, river and floodplain scientists work at a combination of scales, but most of the time, and regardless of the combination of scales, the maximum extent is often bounded by a single river, which in turn has limited the way we developed general theories and concepts on floodplains (e.g. [28]).

7.1.1 What Has Changed with HRT Point-Clouds for Geomorphologists

There are several main changes that have occurred with HRT point-cloud technologies for geomorphologists. The main change maybe comes from the funding, time and skills balance between data acquisition, equipment, data processing and data storage. The idea of the geomorphologist wondering the hills with a pencil and a notebook is now long gone, and the planning of field surveys is an increasing important logistic exercise, with the transport and deployment of UAVs, terrestrial LiDAR or the order of airborne LiDAR surveys often limited to specialized groups and companies. As those tools can generate formidable quality and resolution data, they are not free of original error and they are very voluminous, so that the data processing and information extraction process are costly in equipment and also time-consuming, before geomorphological information can be used.

One of the negative points that has arisen from the HRT point-cloud technologies is the increased disparity that arises between wealthy universities, researchers and countries that can provide a large research envelope to scientists and the rest of the world and the rest of universities. The university where the author works for instance does not have such a fund, and the research funding landscape is not conducive to the acquisition of expensive laser equipment, for instance, as funders prefer providing money for experiments, field expenses and human resources notably. This situation fuels the reliance of countries and researchers "said of the South" on research organizations and groups located in economically wealthy regions of the world. This situation in turn feeds in the gold-rush attitude seen in the disaster risk field [9] and eventually "cheap" research with a post-colonial view of the world [11]. Several years ago, as I was part of a European research group working with Indonesian universities on a major disaster on Java Island, Indonesia, I was stunned—and could not keep it as an untold tale—by the attitude of some of my European colleagues telling our Indonesian counterparts that they should let the European install their own sensors in boreholes that the Indonesian team had dug, because the European knew what they were doing was certainly the one strike too many that helped me realized that even physical sciences had to start include behaviour and ethical guidelines. Those concerns had to extend beyond the realms of social sciences and humanities.

If such attitude has started to be dealt with through ethics committees (at the AGU or the EGU for instance) and researchers have mobilized themselves through manifestos for instance (https://www.ipetitions.com/petition/power-prestige-forgotten-values-a-disaster), the "old-boys club" of geosciences is certainly still in need of change. And the reliance on expensive technologies is not helping ethical and equality progress in geosciences. It only widened the gap. Even in the field of SfM-MVS applied to geomorphology, which should be the low-cost option for everyone, there has been a lock down of research and publications on the topic from the wealthiest countries and institutes in the world (Switzerland, UK, Germany, USA), with eventually the development of protocols, which is certainly a necessary advance for geomorphology as a pure science, but not so much in terms of equity when it becomes compulsory for being published. This is however not a right or wrong question, the development of a protocol to compare dataset is essential for geomorphology to be considered a "serious science" and continue its transition from a field-base discipline with space for subjectivity, to a more numerical and quantitative research discipline. But should we let geomorphology become a more polarized science that is more a reflection of access to funding than ideas? Indeed, there are tools of pointcloud geomorphology, such as SfM-MVS, which are helping science spread more broadly among the different scientific realities (economic, political …). This is because the technology does not require large amount of financial investments, and it can act as a bond between scientists, but also as a tool to develop a more equal system. For better or for worse, the publishing companies that hold on most of the research are the shadow of the same groups that decide on the university ranking worldwide, and this is mostly a political and financial games, leaving out of the run numerous "other universities" with at least as much merit as the "big names". This cycle is completed with the universities pushing their researchers towards more publications, in order to climb the ladder of university fame, and this same number of publications is the driver of ever-increasing database access prices, be it Web of Science or Scopus. All in all, it is a very well-thought business model that feeds itself almost automatically, but is it a good model for intellectual valuation, this is not so certain. And it is certainly a system fuelling post-colonial neo-capitalist disparities. As an academic, in the field of geomorphology—or any other field—one of our missions should be the fair development of geomorphology in countries and at institutions where it is relevant—and the impact of climate change, anthropogenic pressures make geomorphology relevant virtually anywhere on the planet. It should not favour the concentration of academic power in a few limited hands that have theirs on the money already.

At the opposite end of the spectrum where we find SfM-MVS, which is–I believe–a democratic tool, one can find laser scanner techniques. Whether it is a TLS or an ALS, only limited individuals and countries can have access to such tools and technologies for scientific research, and one may want to ask whether it is a real benefit or not for a scientific discipline to open such wide gaps of accessibility to a given tool. Furthermore, there is more than just who owns and who does not own a LiDAR instrument (ALS or TLS), the appearance of the tool invalidates the previously used methods. A quarter of a century ago, you could go in the field on a volcano or anywhere else with a laser pointer taking cross-sections in a gully combined with one single GPS point, and you could present your data with an absolute error of one to several metres, and it was accompanied with various descriptions of the topography and the shapes of the landforms. I am not crying about old time, but I am crying about the gap that has been growing between researchers in so-called rich countries and those in less wealthy countries. The absence of such a tool is now also baring their access to publication. If you don't have an expensive GNSS-RTK or GNSS-PPK or a TLS and other tools to compare your SfM-MVS pointcloud to provide an error and other accuracy and precision data that come from engineering, then your work is disregarded and not published. I am not advocating that we must publish any research, with eventually no control on the data, what I am pointing at are the disparities that this creates.

Of course, researchers from the north will answer that they can collaborate and bring their equipment, but in Indonesia, for example, where I have worked for several decades now, this collaboration is often forced and is under the constraint that without the expensive instrument, no publication will be possible. It even reached a point where the government had to edict new rules to stop all the Indiana Jones to come and steal whatever they could.

7.1.2 What Are Geomorphologists Measuring and Not Measuring with HRT

HRT is providing unprecedented precision and accuracy, but this does not always translate to high accuracy or precision in geomorphology, because the landform is more than just its surface—everybody is aware of it, you are going to tell me. In other words, the shape of the surface and the changes in the surface have been interpreted by geomorphologists within the framework of erosion, deposition and transport processes of sediments, part of the bed rock with large landslides and other flows like lava flows for instance.

It does not mean that we are in a position where we need to reject all the previous conceptualizations, we just need to reflect on when they are relevant. For instance, the translation of surface HRT measurements to erosion provides reliable results of the erosion of bedrock river gorges, which needs millimetre level of accuracy, and from which the mean-change spans from 0.9 to 5.5 mm over a year [3]. As the material cannot be compressed and that the volume/surface relation remains the same, it is possible to transpose previous concepts of geomorphology to these measurements directly. For a rock-cliff face in Japan, Obanawa and Hayakawa [25] and Hayakawa and Obanawa [14] have shown that the erosion rate could be measured by combining TLS and UAV-based SfM. In such a case, there is no relevant variation in the density nor any change in the void ratio of the cliff material, and once again the measure of change in surface can be interpreted as a change in volume and erosion. The DEM-based of the volume of the material at the foot of the cliff however would not be an accurate measure of the deposited volume, because in between the deposited clasts, voids that did not exist on the cliff face will be created, and because the first block falling on the sand beach at the bottom of the cliff is likely to displace and compress the sand they fall onto. A volume measured as a deposit volume would then be difficult to relate to the cliff eroded volume, and it would be wrong to try calculating sediment budget and notably some amount of material exported from the cliff for instance. I would argue that traditional geomorphology with a few data points could work with the present geomorphological paradigms, but trying to increase the precision and the accuracy of calculated volume by increasing the precision and accuracy of the topographic acquisition method would be a mistake, if one works at the centimetre or the millimetre scale.

Millimetre to centimetre-scale measurement of change in bedrock is actually an important change in the geomorphological paradigm already, as it allows short-term measurement of processes occurring over a very long time period. Traditionally, direct measurement would not be deemed possible, and the geomorphologist would have had to interpret long-term changes in geomorphological features (e.g. knick points upward migration dating over long periods), to then bring back long-term change to instantaneous numerical values of erosion. With HRT point-cloud data, it is now possible to make direct measurement of the present processes, and then eventually compare those instantaneous processes against long-term trends. These new possibilities are closing an important methodological gap that traditional geomorphology was facing.

But this improvement comes with a downfall, the necessity to integrate variables that were discarded until present, because not having any significant impacts on the measurements. One of them is soil and material density and compaction.

7.2 The Need to Integrate the Sub-surface: Soil Density Measurement with SfM-MVS and TLS

7.2.1 Why Soil Density Matters for Geomorphologists and Especially with the Rise of High-Resolution Topography

A lot of the work in geomorphology compares surfaces at different periods and differentiating DEMs, for instance, extracting information on the change in volume to make inferences on erosion and deposition notably. This is a method that worked very well for the previous generation of geomorphologists, but this method only works when you have high level of error and equally large surface variations. For instance, if a surface in a landslide area is measured twice, and you record a drop of 10 or 20 m, and if you have an instrument measuring the surface within, let us say ±1 m accuracy, there is no problem with your method, as long as you propagate the error to the volume. Now, if you use the same thought model, but you attempt to make an erosion measure from two surfaces changing by let us say 5 cm with a model accurate to ±1 mm, I argue that you actually don't know the error, because you are lacking an important dataset: compression, decompression and density of the soil you are measuring. These concepts are common in soil engineering, but they were not necessary—so far—in geomorphology. Now that we attempt to make micro-measurements however data on soil density should make its way to our toolbox. This is particularly true if you are working on topographic changes due to cliff collapse or landslide movement, debris flow, etc. …, because the density at the start and at the end is different (Fig. 7.1). Bulk density is therefore an essential parameter that needs to be further included in geomorphology.

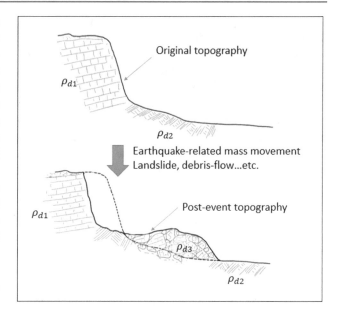

Fig. 7.1 Density change during a wall collapse. The density of materials d1 and d2 and d3 are all different. d1 and d2 differ because they are different formations, and d3 because it includes some erosion underneath the deposit that contributes to the bulk of d3, and also because d1 decompresses to form d3. In other words, if you were to compare d1 and d3, you would certainly see an increase of volume. You could also measure the same volume, but not because they are equal, but because part of it has been removed by other processes. It is therefore essential to include those ideas into your work when using high-resolution topographic data

Bulk density (ρ_b) is the ratio of dry mass to the total volume of soil (e.g. [1, 31]). It is an essential component of soil and sediment analysis, but the bulk density of debris flow and lahar deposits, unwelded ignimbrites (pyroclastic flow deposits) and other coarse-grained mass movement deposits cannot be measured with existing methods. The density of soils and all the sedimentary formations is essential to a broad range of human activities spanning from agriculture to infrastructures and building engineering. It has been used as an estimate of potential soil moisture and porosity (e.g. [18]), as well as soil thermal properties (e.g. [35]). Those relations then change with the median grain size of the soils [21], so that correlation curves cannot be extended through different grain sizes, so that experiments need to be conducted for different grain sizes. The bulk density is also an essential indicator to estimate the maximum shear strength of soil and sediment [32].

7.2.2 Existing Methods for Soil and Sediment Density Measurement

Because both agriculture (and more recently precision agriculture) and infrastructure constructions are mostly occurring on plains, floodplains and coastal areas, the methods that have been developed to measure the bulk density are geared towards material of median grain size ranging from clay to sand and coarse sands. Those methods can be traditionally divided between direct measurement methods and indirect measurement methods (see Al-Shammary et al. [1] for a full review of those methods), with the direct measurement methods being based on a direct measure of both the volume of a sample and the mass of the dry material (Table 7.1). For the direct measurement method, the extraction varies between coring and excavation, while the laboratory analysis relies either on the volume of the extraction cylinder for the coring method [4, 30], while the clod method uses the sample coated in a known volume of paraffin dipped in a liquid of which the displaced volume is measured. For the excavation method, the volume is measured directly in the field, through a calibrated volume of sand filling the excavated whole (sand-cone method) or using an inflatable membrane (the balloon method) of which the amount of liquid in the membrane is measured. The excavation method can be applied when there is a large proportion of fraction above coarse sands, and it even works on slopes [2]—although error is higher [30], but it is only accurate when the material pores are small [8]. Therefore, even if the soil contains gravels and other larger rocks, it is essential to keep a relatively homogenous matrix. Finally, the error margin for these methods is difficult to control, and the results have been shown to be difficult to reproduce [5]. Methods like the sand cone or (sand-replacement method) and the balloon methods respectively need the placement of a metallic plate over the excavated hole and the inflation of a membrane against the excavated hole. It is therefore impossible to know whether the original hole is changing shape or whether the sand or the membrane perfectly fits the excavated surface, and logically, coarse heterometric unconsolidated material provides poor accuracy [19].

The indirect measurement methods (not the object of the present work) include the radiation method, the carbon penetration method and the regression method (Table 7.1). The radiation method is based on the use of the Beer–Lambert's law that relates the distance between the emitted gamma radiation and the received radiation to the density of the material [29]. Another innovative method is based on the pedotransfer to estimate the bulk density (e.g. [24]).

7.2.3　Recent Methodological Developments Using Point-Cloud Technologies

Because there are no method working for all sorts of soils, comparison and overlapping those methods have long been an unanswered problem. In recent years, however, the potential of laser scanning and photogrammetry to measure soil density has started to emerge in the field of soil sciences even if it has not percolated as yet to other geosciences such as geomorphology and geomechanics.

These recent developments include the use of laboratory photogrammetry to determine the volume of a soil clot in the laboratory, where the soil clot <0.8 cm^3 [22], as one would do to obtain the volume and shape of a rock for instance [17]. The comparison of their results (PHM) with the Archimedes volume (AV) provided them with a results of PHM = 1.01 * AV with a R^2 = 0.99. This method is not applicable for the soils and sediment samples we are interested in—loose and unconsolidated—but the photogrammetric framework relies on the same principles. Remote sensing has logically emerged as a complement to the excavation method. Indeed, to obtain the volume of the extracted sample, low-cost laser scanning using the Kinect system has also provided results with an accuracy of R^2 = 0.99 [27]. Similar work has also been conducted using

Table 7.1 Comparison of the direct and indirect ρ_b measurement methods and their range of applicability by characteristic grain sizes' range

		Direct measurement method				Indirect measurement method	
		Coring	Clod method	Excavation method		Radiation method	Carbon penetration
				Sand cone	Balloon		
Grain size characteristics	Clay	o	o	o	o	o	o
	Silt	o	o	o	o	o	o
	Sands	o	x	o	o	o	o
	Gravels	x	x	o	o	o	x
	Coarser	x	x	x	x	x	x
	Coarse mix (gravel and sands)	x	x	x	x	x	x
Error margin control		x	x	x	x	x	x

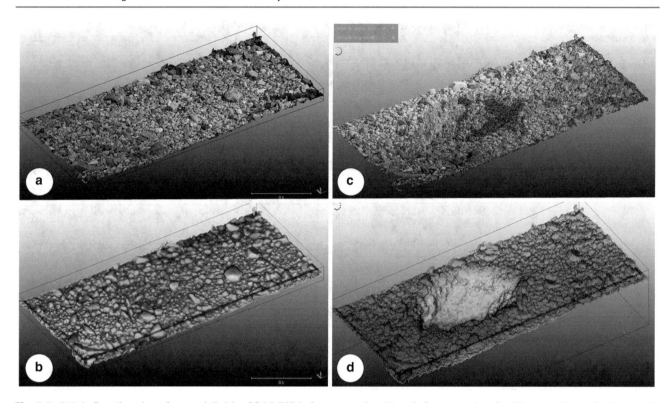

Fig. 7.2 Debris-flow deposit surface modelled by SfM-MVS before excavation (**a**) and after excavation (**b**). The same dataset is shown with relative elevation scales (**b**, **d**) to help show the excavated hole in the ground. The size of the area is 25 cm × 50 cm

traditional photogrammetry using a calibrated camera and a photogrammetry software [2]. Although traditional photogrammetry and laser scanning are attractive methods, both require the acquisition of comparatively costly professional equipment, and specialist knowledge, while SfM-MVS using an "off the shelf" camera and two square rulers gives high-quality results, as I am showing at one sampling location at Volcano Yakedake in Japan (Fig. 7.2).

7.2.4   Your Turn: A Step-by-Step Worked Example

In this section, I describe how I have done such work on material density using laboratory and field data. These data were collected with the following objectives: (1) to test a method relying on the retrieval of the 3D shape of an excavation to measure its volume using SfM-photogrammetry and to compare it to the dry mass of the excavated soil to obtain the density; and (2) to provide an estimate of accuracy of the method based on the photogrammetric error and the characteristics of the material. I would argue that in this example, this is why this method is superior compared to the sand-replacement techniques, for instance, because we have an idea of the error involved with the measurement, while it is difficult to understand how the sand poured into a hole for instance is interacting and lying over another granular

surface: are there any voids or micro-cavities that were not filled or partially filled … etc., and if it is the case, the only way to check it would be to fill the cavities with resin and extract the material as a block. In other words, there aren't practical ways forward. SfM-MVS or laser scanning can provide you with a measure of the error, as well as the location of the error (at the bottom of a hole, on the walls … etc.).

Let's say that you want to measure the density of a debris-flow deposit at different location in the field using SfM-MVS. You will therefore need to have a portable scale, like the squared rulers I have used (Fig. 7.3), and you want to have a precise idea of the geometry of it. In the example, I am using it "as is", but another possibility is to have the same square or rectangle with a set of targets of known location on it. You can then measure precisely the location of those automated targets in the laboratory using a laser scanner or a total station or any other precision instrument and save the data in a file for processing.

Once you have the precise location of the different targets, or a set of known points on your squared ruler, you first want to take sets of photographs for SfM-MVS, in order to measure the potential error in different light environments, indoor, outdoor, in the shade and also taking photographs in different ways. Because the target you are investigating is a "hole" and not a surface plane, you want to take the photographs with a rotation movement, adjusting the roll and

Fig. 7.3 Experimental setup with **a** sand-coated plastic moulds, **b** lahar deposits in the Gokurakudani gully at Unzen Volcano; **c** soil formation in remobilized Quaternary ignimbrite in vertical position in the Gokurakudani gully at Unzen Volcano

pitch, so that you can see the walls of the hole as well as the bottom in as many photographs as possible (see the chapter on SfM-MVS to have an idea of how to best take the photographs). Although authors writing about SfM-MVS provide hints and suggestions on how to take the photographs, this part is more a craft than a given science, and it is most likely that you will better understand how to take the photographs with experience. Once you have your dataset and that you have extracted your pointcloud, you need to measure the exactitude of your measurement. In the present case, I was not only interested in the error measurement from the SfM-MVS process solely but also how the calculated volumes compare to the "real volume", and the best approximation was to compare the real volume of water contained by the excavation to the size of the holes that were modelled (Fig. 7.4).

Using this method in the field or in the laboratory, you can then with high precision, with only a field battery-operated weighing machine and a camera and two rulers (extremely low cost), measure the density of deposits at multiple locations. If you have for instance a splay deposit in a floodplain, a deposit from a tsunami or a fan or a debris-flow cone, you can use this method to measure the bulk density of the deposit at different locations.

If the deposit is not too thick, this simple method allows you to solve the issue of the deposits being of different density compared to the material they originated from.

Imagine that you know a location where the material from a wall is likely to move soon, you can then use one pointcloud generation method to create a pre-event topography and use the method mentioned last to estimate the density of the material (providing that there isn't too much variability

in the material); then, you wait for the event to occur. You take measurements again of the deposits and of the notch left at the original position of the material. You can then measure the density of the deposit, but also the density on the "sliding plane" if it is a small slump you are looking at. Calculating the volume from the notch and the volume on the ground, coupled with the density measurements you would have made, you can then estimate the amount of erosion underneath the slide if you have a discrete event, or this in combination with the amount of material that may have been washed away by other processes. If the event is of consequent size, with a runoff of several metres to hundreds 6636 of metres (or more), you can then the density change along the deposit, and then try to make inferences on the deposition process or you can also try building a more spatially accurate bulk density model.

Pointcloud technology has thus improved our previous measurements and calculations, but it is also providing us with the chance to extend the realm of geomorphology, notably with the potential for accurately measuring density change, compaction change and thus the energy that created the landforms.

Earlier in this chapter, I also wrote that pointcloud technologies had modified our scientific behaviour, notably due to the issues of access and equality, equity of access to technology.

I believe that it has also modified our behaviour in front of the geomorphological fact (landform, process ...). We are now measuring it in the field anymore, we are not doing the conscious act of reflected measure and choice in the field. The data collection step has become an extension of the reflection that will occur in front of the computer. The

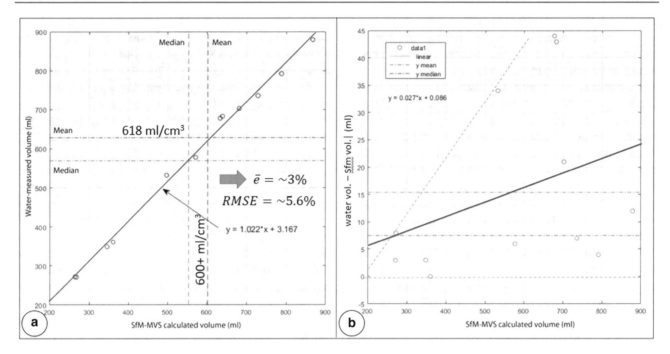

Fig. 7.4 Laboratory comparison of the volume calculated using SfM-MVS and the volume measured from a water-replacement method, with **a** the relation between the two variables (please note that the experimental "hole" needs to be perfectly impermeable to yield significant result, and it is not applicable at all in the field for real data); and with **b** the residuals. The measure of the error is about 3% and the RMSE 5.6%

humanity in the field has been taken for automation and for the multitude. The decision steps are now done in front of our point-cloud processing software, in front of the GIS tools we use to create information from the data. Before point-cloud technology, the scientific act was occurring in the field with a reflection of what is being collected. This has changed as well, and this is what we are discussing in the next paragraphs, and this is—in my view—where the real paradigm shift has occurred in geomorphology.

7.3   Is There a Paradigm Shift, and If so What Is It?

This section is arguably one of the most essential parts of this book, although I must admit that it is the chapter—I believe—which brings more questions than answers and those questions are ones I don't have answers to, and most of us may have no interest in those. But, even those questions do not draw crowds on the streets, this is maybe an exciting part because the existence of these questions means that there is still plenty of research to be done in the field of geomorphology and applied pointcloud technology.

Indeed, I feel compelled to ask whether point-cloud-based HRT is really a game changer or a paradigm shift, or none of the two. Although the concept of "paradigm shift" is often boldly stated in the literature (and I must admit I have written so myself), what does it mean, and for what field?

Geomorphology being the science that investigates the shape of landforms and the processes that build and destroy, modify and reshape them, and how the two relate to one another, are HRT point-clouds a real game changer? Arguably, HRT point-clouds have made geomorphology shine in recent years with a huge number of applications from natural hazards, to civil engineering and other geographies as described in the early chapter of point-clouds applied to geomorphology.

Point-cloud-based geomorphology has emerged as a high-resolution topography tool, used extensively in geomorphology, but the tool by itself is only a topographic measurement at higher resolution. For the geomorphologist, to make full use of this new tool, it is important to integrate it with a knowledge of the sub-surface volume change, depending on applied stress or consolidation over time, etc. Without such consideration, measures of the topography are very difficult to interpret.

In the same way that geomorphology has integrated methods from civil engineering, statistics and physics in the past—for point-cloud technology, the imported methods are laser scanning engineering and geodesy as well as computer vision SfM-MVS—the author proposes to turn once more to civil engineering where the soil (as an engineering soil) needs to be considered for its changes in volume and for the sub-surface transfer of material that affect foundation settlement notably. Even geodetic methods need to be precise to the millimetre when constructing a metal tower or a casted concrete building, but in geomorphology, the measure of one rock could be reduced to its three main axis and considerations of roundness for instance. Thirty years ago, nobody would never have thought of measuring every single gravel on a point bar, for instance, and even more so with several hundred to thousands of measurements per single gravel. Because of the scales at which geomorphology has been operating historically, and because geomorphology works in direct contact with the complexity of nature, micro-variations of surface have often fallen in the box of error or unwanted variability in the dataset. With the advent of HRT point-cloud technology, the geomorphologist is now invited to glorify these small-scale variations and make sense of it.

Furthermore, in geomorphology, the notion of a local site investigation is often closer to a small watershed or a river or coastal section, or a small hill, but very little work has historically zoomed on a plot the size of a building footprint. Therefore, for geomorphologists, change in land surface was interpreted as being the results of erosion, deposition or settlements, but always at large horizontal and vertical scale. Let's consider pingos (also named hydrolaccolith or bulgunniakh), a dome-shaped hill developed in permafrost reaching up to 50–70 m high. Traditionally geomorphologists would measure its overall shape, height, size, general volume using some geometric equation or sets of topographic points, but it was not the domain of geomorphology to look at every square centimetre to look at surface roughness or using multiple survey to look at centimetre-scale vertical variations. With point clouds HRT, this conception of geomorphology has changed.

At first, this change in tools and methods did not come with a necessity to a paradigm shift in geomorphology, because it is mostly research in the field of civil and geotechnical engineering as well as computer vision, but now that geomorphologists are using these tools and methods it is essential to adapt the concepts and assumptions we were traditionally doing to remain relevant.

A change in tool and in method, even if it means modifying some of the concepts of geomorphology—integrating concepts of civil engineering to improve geomorphological measurement—is, I would argue, not the heart of the paradigm shift. It is indeed a modification and an evolution, but I think that the change has occurred from another change.

Davisian geomorphology and the concept of landscape cycles are one of the first ideas developed in scientific geomorphology, and from which we have moved away, but which marks the point zero of a progression that we will see through the twentieth and the twenty-first century. And even if geomorphologists are employing further applied mathematics, in the like of Graph Theory, or geostatistics, as it has typically developed through the leadership of German geomorphology, or by integrating civil engineering concepts

and techniques, as it has been the case under the lead of British geomorphology and to some extent US geomorphology, the relation to the geomorphological object (even those recreated in laboratory flumes) has always been a direct one. As I am interested in this geomorphologist—geomorphology relation, let's close our eyes on all the numerical works that are purely modelling and with no direct connection with existing landforms or processes. Let's interest ourselves in what makes the geomorphologist a natural scientist, its relation to its object. And without dwelling in the difficulties left to us by the philosophy of truth and perception, and truth(s), I would invite the reader to consider the role of point-cloud technologies in this relation. When early geomorphologists were going into the field, doing a sketch accompanied by some measurements, and even when those measurements started to multiply with increasingly complex tools, they were all part of a geomorphological thinking process. Even when I took geomorphological classes, some rather traditional teacher would emphasize the importance of sketching in the field, because it was also a thought process that was occurring through the drawing. We are not simply taking a picture with our eyes that some mechanical parts of our brains are transcribing on the paper, we are, instead, making choices, emphasizing elements and representing parts and sections, which we think or assume, will be of importance to understand either the landform or the geomorphological process. When we do a measurement, the same kind of geomorphological reflection occurs. The geomorphologist is producing a measurement based on assumptions and ideas that are enlightened by one's knowledge and understanding of geomorphology. This is the all part of science that opens for assumptions and then measurement to verify or not those assumptions, and then make theories. Whether we are using a sketchbook or GNSS RTK, or a ground penetrating radar to look at the internal structure, we make expert choices, and these choices have implications on what is measured, and what is measured in the field.

I would argue that pointcloud technologies are removing the geomorphologist thinking process from the field. If the sketchbook and the millimetre accuracy points from the GNSS-RTK or PPK stand on common grounds, I would argue that the point-cloud technologies are not. Even if a GNSS station may look more alike a laser scanner than a pencil and a notebook, this is—from the perspective of geomorphology—just a lure. Indeed, what the pointcloud technology is offering is the promise to provide high-resolution, high-density and high-accuracy (eventually) 3D/4D recording of a landform—even sometimes of a process when measuring it over time—no matter what. The geomorphologist will make choices to go to one site or another, but once in the field, the drone flight or the laser scanning process can be conducted by a surveyor or anybody who can manipulate the tool, but it does not have to be a geomorphologist. Even if the GNSS is not operated by a geomorphologist, her or his opinion of where to take the measurements, for how long, how many times, at which frequency, is all questions she or he needs to be able to answer to study a given geomorphological process or a landform. With laser scanner, it is not the case. We just take it all, and as much as possible.

This is when the geomorphology disappears from the field.

Where does it go?
In the laboratory.

Geomorphology is now happening in front of a computer. The geomorphologist has—after different processing steps—now in front of her or him a representation of the landform or the landform evolution, with more points that one could ever count or dream of, and from the screen, we just have to pick the data that are needed. Do we need a river cross-section? Just draw a line across the pointcloud to extract the points in the vicinity of the line. Do we need the longitudinal profile of a gully, more or less the same ... but the geomorphologist does not do it in the field, she or he does not think about the field, it is now about this virtual environment?

With pointcloud technology, geomorphology is changing from field science to laboratory-based science, to even virtual environment science. This alone is the major shift in geomorphology that pointcloud technology has brought. It has moved the thought process from the field to the "laboratory"; it has removed the geomorphologist from the landscape. Forget your boots and your compass, and put on slippers in front of your screen.

Of course, geomorphology is more than sets of topographic measurements, and there are other operations that still need performing in the field, but arguably the main part of the work, the very relation of the geomorphologist to the landform and the process, is now occurring through the computer screen.

7.4 Pointclouds for Increased Spatial and Temporal Density, as Well as Historical-Time Geomorphology

7.4.1 Bridging Further the Discrete–Continuous Gap

Although I have been using pointcloud technology for over a decade now, I am always wondering what the end game is. Indeed, we have a denser dataset, we can work on gravels and their orientations and a lot of details that were not accessible before, but what does it really means, and what is the limit of this quest for higher density data. Why do I write

dense data and not more accurate or more precise data? Well, this is because it is the increased density that calls for more precision and then accuracy. Indeed, if we measure one point on a rock, we know that variation will occur around that point and that it is representative to some extent, but if we start to have multiple points on the rock, for the data to make sense, we need to make sure that the point data is comparable to the point data around it. We also need to define an added value to these extra-points, to the denser points. And added value there is, without a doubt. But the added value appears in the form of new questions, related to the new scale data being generated, and at a broader scale, the higher density does not help us see or understand a landform better. If you have Mt Fuji characterized by 1 million points on its surface, you can see the shape of it, calculate the slopes and define it as an almost perfect stratovolcano. If you increase the point numbers to 10 billion points for instance. What refinement can you do on the structure itself? You are still seeing a stratovolcano. The new dataset is thus not helping you solve older questions, it is generating its own questions and allowing the scientist to ask questions on smaller, detailed structures. Pointcloud technologies are therefore not really helping answering old questions they are offering a tool to ask new questions in geomorphology. Those questions can now be asked at a much smaller scale, so that at the feature scale (let say we keep our example of Mt Fuji), the feature looks now continuous, the eye of the geomorphologist does not need to interpolate data, the dataset can look like what we have in front of us in reality. The experience of the geomorphological data is now the domain of the landscape experience. There is no need for specialist knowledge to look at the data, when previously the interpolation between one point and another needed a specialist knowledge understanding the limitation of interpolation, the meaning of it ... etc. At the feature scale, statistical estimation is now replaced with collected data. In some ways, it bridges the gap between perception of a continuity (there is no continuity, every atom is separated even if it escapes our eyes) and the data that now also offers the impression, the illusion of continuity.

To an asymptotic movement towards imagined spatial continuity can be added a similar movement towards temporal continuity.

Indeed, HRT pointclouds are allowing scientists to ask new questions that require the collection of data on a daily, weekly or monthly basis. For instance, we can collect weekly topographic data from UAV (let us say an ALS mounted on a UAV) on sand movement on a dune or on the accumulation of guano on a volcanic island due to bird droppings. Keeping with this example, the field of zoogeomorphology can then extend its reach beyond what was thought to be possible before, both thanks to the spatial and temporal increased resolution. Pointcloud technologies are, thus, not only a spatial evolution (or revolution if you agree with the paradigm shift I wrote about in the previous paragraphs) but also a temporal data evolution. We do not need to wait for the yearly governmental flight, and this in turn the high-temporal return data that is now collected from satellites as well. For the geomorphologist, it also has deep implications on the comprehension of processes. Very often we have a dataset before and after an event with sometimes weeks to months between the event itself and the data acquisition, making it difficult to measure the effect of compression of a deposit over time, or its partial remobilization by secondary processes. HRT pointclouds collected from small UAV platforms especially allow the geomorphologist to bridge this gap as well and discover "other" processes that were not visible at a lower temporal resolution as well. And as those secondary processes are usually at a "smaller scale", this is when the HRT also plays an essential role. It is the combination between the two, or high temporal and spatial resolution (HTSR) data or topography (HTSRT), which provides the geomorphologist with the tools to ask new questions and better understand the geomorphic processes.

7.4.2 Giving a Second Life to Historical Aerial Photographs

Furthermore, the possibility to offer a second life to historical aerial photographs has certainly been one of the triggers that had me starting to work with SfM-MVS. In 2011, the author was presenting work looking at the long-term historical evolution of the Tamagawa River in Tokyo, then in 2012, volcanic morphological change over the historical period and more recently at Unzen Volcano [13] and Sakurajima Volcano [10].

This method has now started to emerge in countries where a historical record of high-quality aerial photographs exists. In Canada, Deane et al. [7] have used HRT point-cloud to detect landslide activity and related topographic change. The authors have used photographs at 1:20,000–1:30,000 for the period of 1945–2012 with ~10 years intervals between each series. Comparing LiDAR data and 1982 historical aerial photographs of the "Chin Coulee", for instance, they reported local changes of −5 to +5 m (see Fig. 6 on the 6th page of the paper). The difficulty to work with such photographs, however, is often the lack of overlap between each photograph, because when the photographs were taken, it was not with the aim of doing SfM-MVS calculation from them. Deane et al. [7] reports for their different sets of photographs overlaps of 2–5 only for the years 1945, 1960, 1970, 1982, 1993, 1999 and 2012, which in turn only allows for working from large-scale variations as the error may remain important. Notably Gomez et al. [12] have reported vertical error that can be in

the range of 10 m for densely vegetated areas, where GCPs and location of known altitude are sparse. This is due to the low dpi of the imagery used and also to the dense vegetation cover on mountainous areas. Results are substantially better in rural, urban and semi-urban areas, where GCPs can be multiplied.

SfM-MVS can thus be used as a historical time machine for geomorphologist, allowing us to rebuild past geomorphologies that have been modified either through anthropogenic actions or by natural processes. In other words, the data generated yesterday can be revisited with today's tools in such a way that data that information that could not be extracted 50 or 60 years ago can now be generated. In some ways, we can develop a better understanding of the topography and geomorphology than the contemporary of this geomorphology.

This element is also the origin of a profound change in the approach the geomorphologist can have to the landforms and processes. In other words, you needed to have sets of dedicated data and results to be able to investigate the evolution of landforms over time, for instance, but with SfM-MVS, we can use historical aerial photograph to work on landforms that may not be contemporary anymore, and which were not measured, and for which no existing data is available. SfM-MVS with historical photographs can create this past; it can create the measurements that were not taken at the time the landform or the process occurred, providing that images of it are existing.

Conclusion

Pointcloud technology has definitely a technical component for the geomorphologist, and anyone else using and producing them, but there is also a more fundamental to the emergence of this technology, in terms of what geomorphology interrogates, and what are the possible scales of the questions the geomorphologist can ask. Then emerges the question of how these changes are modifying the scientific "percepts" and how the geomorphologist interacts with its object of research. Finally, the space and time scales that pointcloud technologies have made much finer are now also extended towards the past, with the opening of new possibilities in historical geomorphology, starting in the twentieth century.

References and Suggested Readings

1. Al-Shammary A, Kouzani AZ, Kaynak A, Khoo SY, Norton M, Gates W (2018) Soil bulk density estimation methods: a review. Pedosphere 28:581–596. https://doi.org/10.1016/S1002-0160(18)60034-7
2. Bauer T, Strauss P, Murer E (2014) A photogrammetric method for calculating soil bulk density. J Plant Nutr Soil Sci 177:469–499. https://doi.org/10.1002/jpln.201400010

3. Beer AR, Turowski JM, Kirchner JW (2017) Spatial patterns of erosion in a bedrock gorge. J Geophys Res Earth Surf 191–214. https://doi.org/10.1002/2016JF003850

4. Casanova M, Tapia E, Seguel O, Salazar O (2016) Direct measurement and prediction of bulk density on alluvial soils of central Chile. Chilean J Agric Res 76:105–113. https://doi.org/10.4067/S0718-58392016000100015

5. Chai H, He NP (2016) Evaluation of soil bulk density in Chinese terrestrial ecosystems for determination of soil carbon storage on a regional scale. Acta Ecol Sin 36:3903–3910 (in Chinese). https://doi.org/10.5846/stxb201411222312

6. Cwiakala P, Gruszczynski W, Stoch T, Puniach E, Mrochen D, Matwij W, Matwij K, Nedzka M, Sopata P, Wojcik A (2020) UAV applications for determination of land deformations caused by underground mining. Remote Sens 12:1733, 1–25. https://doi.org/10.3390/rs12111733

7. Deane E, Macciotta R, Hendry MT, Grapel C, Skirrow R (2020) Leveraging historical aerial photographs and digital photogrammetry techniques for landslide investigation—a practical perspective. Landslide 67:1–8. https://doi.org/10.1007/s10346-020-01437-z

8. Frisbie JA, Graham RC, Lee BD (2014) A plaster cast method for determining soil bulk density. Soil Sci 179:103–106. https://doi.org/10.1097/SS.0000000000000044

9. Gaillard JC, Gomez C (2015) Post-disaster research: is there gold worth the rush? Jamba J Disaster Risk Stud 7(1). https://doi.org/10.4102/jamba.v7i1.120

10. Gomez C (2014) Digital photogrammetry and GIS-based analysis of the bio-geomorphological evolution of Sakurajima Volcano, diachronic analysis from 1947 to 2006. J Volcanol Geotherm Res 280:1–13

11. Gomez C, Hart DE (2013) Disaster gold rushes, sophisms and academic neo-colonialism: comments on 'earthquake disaster and resilience in the global north'. Geogr J 179:1272–1277

12. Gomez C, Hayakawa Y, Obanawa H (2015) A study of Japanese landscapes using structure from motion derived DSMs and DEMs based on historical aerial photographs: new opportunities for vegetation monitoring and diachronic geomorphology. Geomorphology 242:11–20

13. Gomez C, Shinohara Y, Tsunetaka H, Hotta N, Bradak B, Sakai Y (2021) Twenty-five years of geomorphological evolution in the Gokurakudani gully (Unzen Volcano): topography, subsurface geophysics and sediment analysis. Geosciences 11. https://doi.org/10.3390/geosciences11110457

14. Hayakawa YS, Obanawa H (2020) Volumetric change detection in bedrock coastal cliff using terrestrial laser scanning and UAS-based SfM. Sensors 20:3403. https://doi.org/10.3390/s20123403

15. Hofland B, Diamantidou E, van Steeg P, Meys P (2015) Wave runup and wave overtopping measurements using a laser scanner. Coast Eng 106:20–29

16. Huttett RJ (2011) Fundamentals of geomorphology, 2nd edn. Routledge

17. James MR, Robson S (2012) Straightforward reconstruction of 3D surfaces and topography with a camera: accuracy and geoscience application. J Geophys Res 117:F03017. https://doi.org/10.1029/2011JF002289

18. Lu YL, Liu XN, Zhang M, Heitman J, Horton R, Ren TS (2017) Thermo-time domain reflectometry method: advances in monitoring in situ soil bulk density. Soil Sci Soc Am J. https://doi.org/10.2136/msa2015.0031

19. Ma YY, Lei TW, Zhang XP, Chen YX (2013) Volume replacement method for direct measurement of soil moisture and bulk density. Trans Chin Soc Agric Eng 29:86–93 (in Chinese). https://doi.org/10.3969/j.issn.1002-6819.2013.09.012

20. Mao L, Ravazzolo D, Bertoldi W (2020) The role of vegetation and large wood on the topographic characteristics of braided river systems. Geomorphology 107299, 1–11

21. Martin MA, Reyes M, Taguas FJ (2017) Estimating soil bulk density with information metrics of soil texture. Geoderma 287:66–70. https://doi.org/10.1016/j.geoderma.2016.09.008

22. Moret-Fernandez D, Latorre B, Pena C, Gonzales-Cebollada C, Lopez M (2016) Applicability of the photogrammetry technique to determine the volume and the bulk density of small soil aggregates. Soil Res 54:354–359. https://doi.org/10.1071/SR15163

23. Morgan JA, Brogan DJ, Nelson PA (2017) Application of structure-from-motion photogrammetry in laboratory flumes. Geomorphology 276:125–143

24. Nanko K, Ugawa S, Hashimoto S, Imaya A, Kobayashi M, Sakai H, Ishizuka S, Miura S, Tanaka N, Takahashi M, Kaneko S (2014) A pedotransfer function for estimating bulk density of forest soil in Japan affected by volcanic ash. Geoderma 213:36–45. https://doi.org/10.1016/j.geoderma.2013.07.025

25. Obanawa H, Hayakawa YS (2018) Variations in volumetric erosion rates of bedrock cliffs on a small inaccessible coastal island determined using measurements by an unmanned aerial vehicle with structure-from-motion and terrestrial laser scanning. Prog Earth Planet Sci 5(33):1–10

26. Robinson SE, Bohon W, Kleber EJ, Arrowsmith JR, Crosby CJ (2017) Applications of high-resolution topography in earth science education. Geosphere 13(6):1887–1900

27. Scanlan CA, Rahmani H, Bowles R, Bennamoun M (2018) Three-dimensional scanning for measurement of bulk density in gravelly soils. Soil Use Manage 1–8. https://doi.org/10.1111/sum.12426

28. Scown MW, Thoms MC, De Jager NR (2015) Floodplain complexity and surface metrics: influences of scale and geomorphology. Geomorphology 245:102–116

29. Smith KA (2000) Soil and environmental analysis: physical methods, revised and expanded. CRC Press, New York

30. Vanguelova EI, Bonifacio E, De Vos B, Hoosbeek MR, Berger TW, Vesterdal L, Armolaitis K, Celi L, Dinca L, Kjonaas OJ, Pavlenda P, Pumpanen J, Puttsepp U, Reidy B, Simoncic P, Tobin B, Zhiyanski M (2016) Sources of errors and uncertainties in the assessment of forest soil carbon stocks at different scales—review and recommendations. Environ Monit Assess 188:630. https://doi.org/10.1007/s10661-016-5608-5

31. Walter K, Don A, Tiemeyer B, Freibauer A (2016) Determining soil bulk density for carbon stock calculations: a systematic method comparison. Soil Sci Soc Am J 80:579–591. https://doi.org/10.2136/sssaj2015.11.0407

32. Wilson MG, Sasal MC, Caviglia OP (2013) Critical bulk density for a Mollisol and a Vertisol using least limiting water range: effect on early wheat growth. Geoderma 192:354–361. https://doi.org/10.1016/j.geoderma.2012.05.021

33. Xu C, Huang Z, Yao Y (2019) A wave-flume study of scour at a pile breakwater: solitary waves. Appl Ocean Res 82:89–108

34. Yang C-J, Jen C-H, Cheng Y-C, Lin J-C (2021) Quantification of mudcracks-driven erosion using terrestrial laser scanning in laboratory runoff experiment. Geomorphology 375:107527, 1–10

35. Zhang T, Gai GJ, Liu SY, Puppala AJ (2017) Investigation on thermal characteristics and prediction models of soils. Int J Heat Mass Transfer 106:1074–1086. https://doi.org/10.1016/j.ijheatmasstransfer.2016.10.084

List of Worked Case Studies

Chapter 1

(In this introductory chapter, there aren't any worked example.)

Chapter 2

2.2.3 Example of a Step-by-Step Workflow with Agisoft Metashape-Pro

In this example, I provide you with a simple tutorial on how to use the proprietary software Agisoft Metashape-Pro (the choice is motivated by its ubiquitous use through geoscience, and it is most likely that your laboratory will have a licence of the software). The example explains the different steps from importing your photographs to adding GCPs and generating derivatives such as the DEM or the DSM and the aerial photographs. There are other options in the software, but I am presenting you with all the necessary steps towards generating a geographic dataset.

Chapter 3

In this chapter, snypet of codes and different elements are weaved into the chapter, as it has an emphasis on the technical aspects, instead of presenting examples separately solely. Then, I do encourage you to read through the chapter to find elements that might interest you.

3.3 Example to Extract Data on a Grid from a Point-Cloud with R and LidR

Using R and LidR, I present you with methods to grid your LiDAR or eventual SfM-MVS dataset, with different grid size, for the Cuckmere Haven in the UK. The example also shows the importance of choosing the right gridding steps and methods depending on what you want to show (i.e. there is no point having to deal with too much detail if it is of no interest to your work).

Chapter 4

4.4 Worked Example with R: A Meander on the Alabama River

This example is rather extensive and I explain how to extract LiDAR data from the USGS database for a given meander of the Alabama River, then how to display the different information in the dataset (such as the swath number, the altitude, the return intensity, etc.). Then using the classified point-cloud, I extract the ground level, which reveals all the micro-morphology of the floodplain and the point-bars. I also show how you can choose to save some parts of the pointcloud (like the ground only) and how to extract transects out of the datasets, still in R. Although there are other tools that can do this same work (and I am presenting the open-source ones in this book as well), R also allows you to perform a very broad variety of statistical tests on your dataset, and being able to handle the LiDAR data inside R will save you a lot of time.

4.5 Worked Coastal Example with R

In this example, we work on coastal data including sea surface data, and using R and the LidR package, we are extracting spatial portion of the dataset, extracting outcrops.

Chapter 5

5.3.1 Road-Side Slope Monitoring Using Low-Cost SLAM-LiDAR Sensor and SfM-MVS

In this example, we are comparing two point-clouds, one acquired from SLAM-LiDAR and one using SfM-MVS. This example shows how you can follow erosional features at the local scale without a world-coordinate system if you are only interested in the local change. It is also a useful method under dense forest cover, where other airborne methods are difficult

© Springer Nature Switzerland AG 2022
C. Gomez, *Point Cloud Technologies for Geomorphologists*,
Springer Textbooks in Earth Sciences, Geography and Environment,
https://doi.org/10.1007/978-3-031-10977-5_1

to apply. In this example we use R and the LidR library to process the data, and then the CloudCompare Open-source software using the M3C2 algorithm to compare the two successive pointclouds.

5.3.2 Aligning a Debris-Flow Fan Lidar-Based DEM and a SfM-MVS Point-Cloud

This example presents two datasets in mountain forest environments, one obtained from LiDAR before a debris flow occurred and generated a cone deposit, and one created afterwards, with UAV photographs and SfM-MVS. The UAV has on-board a simple GNSS, so that a roughly georeferenced pointcloud was created. The example then compared 213 points that did not change with the debris flow to compare the two datasets. Then I show the data after GCP use, from which the residual error can be deduced (about 40 cm).

5.3.3 Detrending the Topography to Extract Small-Scale Features and Trees in R

In the previous example, trees are one of the elements that limit the number of GCPs that can be used, so LiDAR data should be preferred to SfM-MVS in highly vegetated areas. Still, you need to be able to remove some of the vegetation, and although there are numerous methods to do so, as a first step, I always look at the structure of the vegetation, because it can hold some information about the erosive processes. For instance, you may have patches of trees in a natural environment, which are regrowing after an erosive event or portion of the vegetation along the riparian zone that has a given age due to a previous flood or debris flow, etc. In the present case, I use R and the LidR library to eventually identify the tree structure and show the trees independently of the topography, by detrending the dataset.

5.3.4 Using the Trees as a Proxy of Slope and Other Geomorphologic Processes

This example is an extension of an example developed in 5.3.3, except that this time, I am using a function in LidR in order to extract each single tree that has a given height above the ground, so that you can link a growth function for each tree to its species … etc. This can be particularly useful if you work in the field of forestry and you want to compare how tall a given planted tree is by comparing the data to the geomorphology and climatic and weather data.

5.3.5 Micro- and Local Slope Change Example in the Rokko Mountains of Japan Using SfM-MVS

The final example of Chap. 5 is a local-scale feature extraction from the 3D of an outcrop that was generated using SfM-MVS with GCP targets which were measured using a total station. The example does not focus so much on the data acquisition as the first example did, but it shows the different steps of the SfM-MVS process (the emphasis was on other points in the other examples, and for error analysis, please, see the examples of Chap. 6).

Chapter 6
6.3.1 Finding, Loading and Doing Virtual Fieldwork on a US Volcano: Mt. Shasta of California

For education purposes, especially in the face of the COVID-19 crisis, virtual fieldworks have become very popular. In the present case, I have downloaded a set of LiDAR data from the USGS database at Mount Shasta Volcano in California. From the downloaded dataset, I show how you can use the Autodesk software Autodesk-Recap (free student and instructor licences available) to calculate slopes and distances in just a few clicks of the mouse.

6.3.2 Measuring the Depth, Eroded Bulk Volume (Material and Air/Water in the Volume) and Erosion Rate from UAV-Based Photogrammetry at Izu-Oshima Volcano

In this example, I explain how to take aerial photographs from a UAV to generate a 3D model and how to use a set of GCPs to measure the error from the dataset. In the 1986 fallout deposits, a network of gullies has been eroding, and from topographic measure I am showing you how you can estimate the eroded volumes.

6.3.3 SfM-MVS for Laboratory-Scale Volcanic Structure and Forms

This case study explains the procedure to set up a laboratory experiment with an evolving topographic structure, and how to place GCPs around the structure. I also explain what are the main error measurements that can be made and show an example of the spatial distribution of this error.

Chapter 7
7.2.4 Soil Density Calculation from SfM-MVS

In this example, using SfM-MVS measure of a pre-hole and of a dug hole, compared with the bulk mass of the dug-out material, provides certainly the best method to measure the soil density in granular material. This operation can be done at different depths by working by layers. In this example, I explain the method and also explain the accuracy of the method using water-volume equivalents.